U.S. NAVY SEAL SNIPER TRAINING PROGRAM

U.S. Navy

Skyhorse Publishing

Photos on pages 49, 82, 111, 149, 161, 171, 180, 187, 206, 223, 231, 240, and 246
courtesy of Brandon Webb and Glen Doherty

Skyhorse Publishing books may be purchased in bulk at special discounts for sales promotion, corporate gifts, fund-raising, or educational purposes. Special editions can also be created to specifications. For details, contact the Special Sales Department, Skyhorse Publishing, 307 West 36th Street, 11th Floor, New York, NY 10018 or info@skyhorsepublishing.com.

Skyhorse® and Skyhorse Publishing® are registered trademarks of Skyhorse Publishing, Inc.®,
a Delaware corporation.

www.skyhorsepublishing.com

10 9 8 7 6 5 4 3 2 1

Library of Congress Cataloging-in-Publication Data is available on file.
ISBN: 978-1-61608-223-9

Printed in Canada

TABLE OF CONTENTS

TABLE OF CONTENTS

NAVAL SPECIAL WARFARE
BASIC SNIPER TRAINING SYLLABUS

NOTE. The primary mission of the SEAL scout sniper in combat is to support combat operations by delivering percision fire on selected targets from concealed postions. The SEAL sniper also has a secondary mission of gathering information for intelligence purposes. The future combat operations that would most likely involve Naval Special Warfare would be low intensity type combat operations that would employ SEAL personnel in the gathering of information for future military operations or the surgical removal of military targets and personnal with a minimal assult force with no lost of life to civilan personnel, this is the ideal mission profile to employ snipers due to their advanced field skills, marksmanship and their ablity to operate independently in a field environment.

1. HOURLY BREAKDOWN OF 9-WEEK PERIOD OF INSTRUCTION.

HOURS	SUBJECT
40	NAVAL GUN FIRE SUPPORT SCHOOL (LITTLECREEK VA.)
4	ZEROING.
20	UNKNOWN DISTANCE FIRING.
66.5	STATIONARY TARGET FIRING (M-14/BOLT RIFLE).
31	MOVING TARGET FIRING.
8	NIGHT FIRING UNDER ARTIFICIAL ILLUMINATION.
15	SHOOTING TESTS - STATIONARY/MOVING/POPUP TARGETS.
12	COMBAT PISTOL SHOOTING.
12	HELO INSERTIONS/EXTRACTIONS-CALL FOR FIRE (4 APPLICATION EXERCISES- 2 NIGHT/2 DAY)
11.5	EMPLOYMENT/MISSION PLANING RELATED CLASSIES.
28	COMMUNICATIONS INSTRUCTION(6 APPLICATION EXERCISES) LST-5B, AN/PRC-117, PSC-3, AN/PRC-113.
42	MAPPING/AERIALPHOTO INSTRUCTION(6 APPLICATION EXERCISES)
1.5	WRITTEN TESTS.
44	STALKING EXERCISES(11 EXERCISES).
11	RANGE ESTIMATION EXERCISES(11 EXERCISES).

11	OBSERVATION EXERCISES(11EXERCISES).
6	CONCEALMENT EXCERCISES(3 EXERCISES).
10	HIDE CONSTRUCTION EXERCISE(1 EXERCISE).
72	MISSION EXERCISES(3,EACH COVERING A 24-HOUR PERIOD).
16	TACTICAL EXERCISE WITHOUT TROOPS(TEWT) (4EXERCISES).
500	TOTAL HOURS

2. SNIPER PROFICIENCY TRAINING.

The purpose of proficiency training is to enable the qualified SEAL scout sniper to maintain the degree of skill and proficiency to which he was trained. Proficiency training should be conducted on a quarterly in all sniper skills, although special emphasis should be made on marksmanship and stalking. These should be practiced as frequenly as possible. Every effort should be made to maintain sniper proficiency.

Snipers should be requalified each year in all SEAL scout sniper skills. They should also be "quizzed" and/or tested every quarter. Proficiency training should be conducted to the same degree of standards as it was originally taught so not to lose any effectiveness in combat. If a sniper is not retained quarterly in all basic sniper skills, his quality of performance will decrease; therefore, he will not meet the standards of the SEAL scout sniper.

NOTE: SEAL scout snipers must be included, in the sniper roll, in normal SEAL tactical training and in tactical exercises.

INTERNAL SECURITY EMPLOYMENT

INTRODUCTION

1. Gain Attention. Imagine Special Warfare suddenly committed to a peace keeping force such as in Beirut, Lebanon. Or, imagine being committed to preserve the peace and protect innocent lives and property in an urban environment such as Detroit or Watts during a "Big City" riot. What is the role of the sniper? Is the sniper a valid weapon for employment in situations like this?

2. The answer is most emphatically, yes!! We have only to look around us to see examples of how effective the sniper can be in this type of situation. Probably the best examples available to us are two recent British involvements: Aden and Northern Ireland. In both cases the sniper has played a significant role in the successful British peace keeping efforts. Remember, that one of the key principles of crowd control/peace keeping is the use of only minimum force. The sniper with his selective target identification and engagement with that one well aimed shot is one of the best examples of the use of minimum force.

3. Purpose

a. Purpose. To provide the student with the general knowledge needed to employ a sniper section in internal security type environments.

b. Main Ideas. To explain the sniper's role in:

(1) Urban guerrilla operations.
(2) Hostage situations.

4. Training Objectives. Upon completion of this period of instruction, the student will be able to:

a. Employ a Seal sniper section in either sniper cordon, periphery, O.P. or ambush operations.

b. Construct and occupy an urban O. P.

c. Obtain and use special equipment needed for internal security operations.

d. Employ a Seal sniper section in a hostage situation.

e. Select a hostage situation firing position taking into consideration the accuracy requirements and effects of glass on the bullet.

BODY

1. Urban Guerrilla Warfare

a. General. The role of the sniper in an urban guerrilla environment is to dominate the area of operations by delivery of selective, aimed fire against specific targets as authorized by local commanders. Usually this authorization only comes when such targets are about to employ firearms or other lethal weapons against the peace keeping force or innocent civilians. The sniper's other role, and almost equally important, is the gathering and reporting of intelligence.

b. Tasks. Within the above role, some specific tasks which may be assigned include:

(1) When authorized by local commanders, engaging dissidents/ urban guerrillas when involved in hijacking, kidnapping, holding hostages, etc.

(2) Engaging urban guerrilla snipers as opportunity targets or as part of a deliberate clearance operation.

(3) Covertly occupying concealed positions to observe selected areas.

(4) Recording and reporting all suspicious activity in the area of observation.

(5) Assisting in coordinating the activities of other elements by taking advantage of hidden observation posts.

(6) Providing protection for other elements of the peace keeping force, including fireman, repair crews, etc.

c. Limitations. In urban guerrilla operations there are several limiting factors that snipers would not encounter in a conventional war:

(1) There is no FEBA and therefore no "No Mans Land" in which to operate. Snipers can therefore expect to operate in entirely hostile surroundings in most circumstances.

(2) The enemy is covert, perfectly camouflaged among and totally indistinguishable from the everyday populace that surrounds him.

(3) In areas where confrontation between peace keeping forces and the urban guerrillas takes place, the guerrilla dominates the ground entirely from the point of view of continued presence and observation. Every yard of ground is known to them; it is ground of their own choosing. Anything approximating a conventional stalk to and occupation of, a hide is doomed to failure.

(4) Although the sniper is not subject to the same difficult conditions as he is in conventional war, he is subject to other pressures.

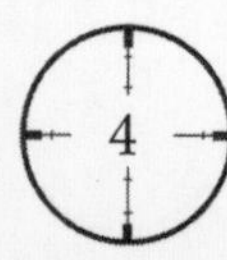

These include not only legal and political restraints but also the requirement to kill or wound without the motivational stimulus normally associated with the battlefield.

(5) Normally in conventional war, the sniper needs no clearance to fire his shot. In urban guerrilla warfare, the sniper must make every effort possible to determine in each case the need to open fire and that it constitutes reasonable/minimum force under circumstances.

d. Methods of Employment

(1) Sniper Cordons/Periphery O. P. 's

(a) The difficulties to be overcome in placing snipers in heavily populated, hostile areas and for them to remain undetected, are considerable. It is not impossible, but it requires a high degree of training, not only on the part of the snipers involved, but also of the supporting troops.

(b) To overcome the difficulties of detection and to maintain security during every day sniping operations, the aim should be to confuse the enemy. The peace keeping forces are greatly helped by the fact that most "trouble areas" are relatively small, usually not more than a few hundred yards in dimension. All can be largely dominated by a considerable number of carefully sited O. P. 's around their peripheries.

(c) The urban guerrilla intelligence network will eventually establish the locations of the various O. P. 's. By constantly changing the O. P.'s which are in current use it is impossible for the terrorist to know exactly which are occupied. However, the areas to be covered by the O.P.'s remain fairly constant and the coordination of arcs of fire and observation must be controlled at a high level, usually battalion. It may be delegated to company level for specific operations.

(d) The number of O.P.'s required to successfully cordon an area is considerable. Hence, the difficulties of sustaining such an operation over a protracted period in the same area should not be under-estimated.

(2) Sniper Ambush

(a) In cases where intelligence is forth coming that a target will be in a specific place at a specific time, a sniper ambush is frequently a better alternative than a more cumbersome cordon operation.

(b) Close reconnaissance is easier than in normal operation as it can be carried out by the sniper as part of a normal patrol without party to its hide undetected. To place snipers in position undetected will require some form of a deception plan. This often takes the form of a routine search operation in at least platoon strength. During the course of the search the snipers position themselves in their hide. They remain in position when the remainder of the force withdraws. This tactic is especially effective when carried out at night.

(c) Once in position the snipers must be prepared to remain for lengthy periods in the closest proximity to the enemy and their sympathizers.

(d) Their security is tenuous at best. Most urban O.P.'s have "dead spots" and this combined with the fact that special ambush positions are frequently out of direct observation by other friendly forces makes

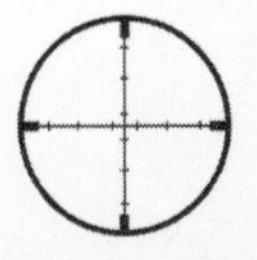

with explosives. The uncertainty about being observed on entry is a constant worry to the snipers. It can and does have a most disquieting effect on the sniper and underlines the need for highly trained men of stable character.

(e) If the ambush position cannot be directly supported from a permanent position, a "back up" force must be placed at immediate notice to extract the snipers after the ambush or in the event of compromise. Normally it must be assumed that after the ambush, the snipers cannot make their exit without assistance. They will be surrounded by large, extremely hostile crowds, consequently the "back up" force must not only be close at hand but also sufficient in size.

c. Urban Sniping Hides/O.P.'s

(1) Selecting the Location. The selection of hides and O.P. positions demand great care. The over-riding requirement of a hide/O.P. position is for it to dominate its area of responsibility.

(a) When selecting a suitable location there is always a tendency to go for height. In an urban operation this can be mistake. The greater the height attained, the more the sniper has to look out over an area and away from his immediate surroundings. For example, if an O.P. were established on the 10th floor of an apartment building, to see a road beneath, the sniper would have to lean out of the window, which does little for the O.P.'s security. The locations of incidents that the sniper might have to deal with are largely unpredictable, but the ranges are usually relatively short. Consequently, an O.P. must aim to cover its immediate surroundings as well as middle and far distances. In residential areas this is rarely possible as O.P.'s are forced off ground floor level by passing pedestrians. But generally it is not advisable to go above the passing pedestrians. But generally it is not advisable to go above the second floor, because to go higher greatly increases the dead space in front of the O.P. This is not a cardinal rule, however. Local conditions, such as being on a bus route, may force the sniper to go higher to avoid direct observation by passengers.

(b) In view of this weakness in local defense of urban O.P.'s, the principles of mutual support between O.P.'s assumes even greater importance. The need for mutual support is another reason for coordination and planning to take place at battalion level.

(c) The following are possible hide/O.P. locations:

(1) Old, derelict buildings. Special attention should be paid to the possibility of encountering booby traps. One proven method of detecting guerrilla booby traps is to notice if the locals (especially children) move in and about the building freely.

(2) Occupied houses. After careful observation of the inhabitants daily routine, snipers can move into occupied homes and establish hides/O.P.s in the basement and attics. This method is used very successfully by the British in Northern Ireland.

(3) Shops.

(4) Schools and Churches. When using these as hide/O.P. locations, the snipers risk possible damage to what might already be strained public relations.

(5) Factories, sheds, garages.

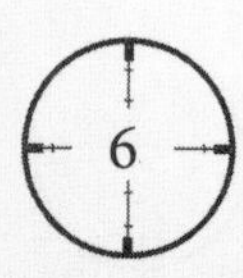

(6) Basements and between floors in buildings. It is possible for the sniper team to locate themselves in these positions although there may be no window or readily usuable firing port available. These locations require the sniper to remove bricks or stone without leaving any noticeable evidence outside of the building. To do this the sniper must carefully measure the width of the mortar around a selected brick/stone. He must then construct a frame exactly the size of the selected brick with the frame edges exactly the size of the surrounding mortar. He then carefully removes the brick from the wall and places it in his frame. The mortar is then crushed and glued to the frame so that it blends perfectly with the untouched mortar still in place. The brick/frame combination is then placed back into the wall. From the outside, nothing appears abnormal, while inside the sniper team has created an extremely difficult to detect firing port. Care must be taken however that when firing from this position dust does not get blown about by muzzle blast and that the brick/frame combination is immediately replaced. Another difficulty encountered with this position is that it offers a very restricted field of view.

(7) Rural areas from which urban areas can be observed.

(d) An ideal hide/O.P. should have the following characteristics:

(1) A secure and quiet approach route. This should, if possible, be free of garbage cans, crumbling walls, barking dogs and other impediments.

(2) A secure entry and exit point. The more obvious and easily accessible entry/exit points are not necessarily the best as their constant use during subsequent relief of sniper teams may more readily lead to compromise.

(3) good arcs of observation. Restricted arcs are inevitable but the greater the arc the better.

(4) Security. These considerations have already been discussed above.

(5) Comfort. This is the lowest priority but never the less important. Uncomfortable observation and firing positions can only be maintained for short periods. If there is no adequate relief from observation, O.P.s can rarely remain effective for more than a few hours.

(2) Manning the O.P./Hide

(a) Before moving into the hide/O.P. the snipers must have the following information:

(1) The exact nature of the mission (i.e. observe, shoot, etc.)
(2) The length of stay.
(3) The local situation.
(4) Procedure and timing for entry.
(5) Emergency evacuation procedures.
(6) Radio procedures.
(7) Movement of any friendly troops.
(8) Procedure and timing for exit.
(9) Any special equipment needed.

(b) The well-tried and understood principle of remaining back from windows and other apertures when in buildings has a marked effect on the manning of O.P.s/hides. The field of view from the back of a room through a window is limited. To enable a worthwhile area to be covered, two or even three men may have to observe at one time from different parts of the room.

(3) Special Equipment for Urban Hides/O.P. The following equipment may be necessary for construction of or use in the urban/O.P.

(a) Pliers. To cut wires.
(b) Glass Cutter. To remove glass from windows.
(c) Suction Cups. To aid in removing glass.
(d) Rubber Headed Hammers. To use in construction of the hide with minimal noise.
(e) Skeleton Keys. To open locked doors.
(f) Pry Bars. To open jammed doors and windows.
(g) Padlocks. To lock doors near hide/O.P. entry and exit points.

2. Hostage Situations

a. General. Snipers and commanding officers must appreciate that even a good, well placed shot may not always result in the instantaneous death of a terrorist. Even the best sniper when armed with the best weapon and bullet combination cannot guarantee the desired results. Even an instantly fatal shot may not prevent the death of a hostage when muscle spasms in the terrorists's body trigger his weapon. As a rule then, the sniper should only be employed when all other means of moving the situation have been exhausted.

b. Accuracy Requirements

(1) The Naval Special Warfare Sniper Rifle is the finest combat sniper weapon in the world. When using the Lake City M118 Match 7.62 mm ammunition it will constantly group to within one minute of angle or one inch at one hundred yards.

(2) Keeping this in mind, consider the size of the target in a hostage situation. Doctors all agree that the only place on a man, where if struck with a bullet instantaneous death will occur, is the head. (Generally, the normal human being will live 8-10 seconds after being shot directly in the heart.) The entire head of a man is a relatively large target measuring approximately 7 inches in diameter. But in order to narrow the odds and be more positive of an instant killing shot the size of the target greatly reduces. The portion of the brain that controls all motor relex actions is located directly behind the eyes and runs generally from ear lobe to ear lobe and is roughly two inches wide. In reality then, the size of the snipers target is two inches not seven inches.

(3) By applying the windage and elevation rule, it is easy to see then that the average Seal sniper cannot and should not attempt to deliver an instantly killing head shot beyond 200 yards. To require him to do so, asks him to do something the rifle and ammunition combination available to him cannot do.

c. Position Selection. Generally the selection of a firing position for a hostage situation is not much different from selecting a firing position for any other form of combat. The same guidelines and rules apply. Remember, the terrain and situation will dictate your choice of firing positions. However, there are several peculiar considerations the sniper must remember:

(1) Although the sniper should only be used as a last resort, he should be moved into his position as early as possible. This will enable him to precisely estimate his ranges, postively identify both the hostages and the terrorist and select alternate firing positions for use if the situation should change.

(2) If the situation should require firing through glass, the sniper should know two things:

(a) That when the Mils ammunition penetrates glass, in most cases the copper jacket is stripped off its lead core and fragments. These fragments will injure or kill should they hit either the hostage or the terrorist. The fragments show no standard pattern but randomly fly in a cone shaped pattern much like shot from a shotgun. The lead core of the bullet does continue to fly in a straight line. Even when the glass is angled to as much as 45° the lead core will not show any signs of deflection. (back 6 feet from the point of impact with the glass).

(b) That when the bullet impacts with the glass, the glass will shatter and explode back into the room. The angle of the bullet impacting with the glass has absolutely no bearing on the direction of the flight of the shattered glass. The shattered glass will always fly perpendicular to the pane of the glass.

d. Command and Control

(1) Once the decision has been made by the commander to employ the sniper, all command and control of his actions should pass to the sniper team leader. At no time should the sniper have to fire on someone's command. He should be given clearance to fire and then he and he alone should decide exactly when.

(2) If more than one sniper team is used to engage one or more hostages it is imperative that the rule above applies to all teams. But it will be necessary for the snipers to communicate with each other. The most reliable method of accomplishing this is to establish a "land line" or TA-312 phone loop much like a gun loop used in artillery battery firing positions. This enables all teams to communicate with all the others without confusion about frequencies, radio procedure, etc.

OPPORTUNITY FOR QUESTIONS AND COMMENTS

SUMMARY

1. Reemphasize. During this period of instruction we discussed urban guerrilla operations and hostage situations. In urban guerrilla operations we outlined the tasks and limitations common to all operations. We then discussed the two methods of employing snipers: (1) sniper cordons/ periphery O.P.s and (2) sniper ambushes. We discussed selecting a position

in an urban area and the most suitable locations for hides/O.P.s. Then we looked at how to man an O.P. and what special equipment you might need to construct and work in it.

In the discussion of hostage situations we examined the accuracy requirements and the position selection considerations common to all terrorist environments. We also discussed the command and co trol procedures for employing snipers in this type of role.

2. Remotivate. Remember, its not outside the realm of possibility that some ay you or someone you've trained could find himself in this type of situation. At that time you'll take the test—let's hope we have no failures, because the political and social repercussions are too great a price to pay for one sniper who didn't prepare himself to put that one round on target.

OCCUPATION
AND
SELECTION OF POSITIONS

INTRODUCTION

1. Gain Attention. Relate story of Russian super-sniper Vassili Zaitsev and German super-sniper Major Konigs at the Battle of Stalingrad. (Excerpts from Enemy at the Gates by William Craig.)

2. Simply stated, the Specwar snipers mission is to see without being seen and to kill without being killed.

3. Purpose

a. Purpose. The purpose of this period of instruction is to provide the student with the knowledge required to select and occupy a position.

b. Main Ideas. The main ideas to be discussed are the following:

(1) Position Selection
(2) Hasty Positions
(3) Position Safety
(4) Actions in Position

4. Training Objectives. Upon completion of this period of instruction the student will:

a. Identify those features which contribute to the selection of a position. i. e. cover, concealment, fields of fire, avenues of approach and withdrawal, etc.

b. Determine, using maps, aerial photos and/or visual reconnaissance, the location of a suitable sniper position.

TRANSITION. To effectively accomplish their mission of supporting combat operations by delivering precision fire on selected targets the sniper team must select a position from which to observe and fire.

BODY

1. Position Selection. The sniper, having decided upon an area of operation, must chose a specific spot from which to operate. The sniper must not forget that a position which appears to him as an obvious and ideal location for a sniper will also appear as such to the enemy. He should avoid the obvious positions and stay away from prominent, readily

identifiable objects and terrain features. (TA) The best position represents an optimum balance between two considerations.

a. It provides maximum fields of observation and fire to the sniper.

b. It provides maximum concealment from enemy observation.

2. Hasty Positions. Due to the limited nature of most sniper missions and the requirement to stalk and kill, the sniper team will in most cases utilize a hasty post. Considering the fundamentals of camouflage and concealment the team can acquire a hasty sniper post in any terrain. (TA) The principle involved when assuming a hasty position is to utilize a maximum of the team's ability to blend with the background or terrain and utilize shadows at all times. Utilizing the proper camouflage techniques, while selecting the proper position from which to observe and shoot, the sniper can effectively preclude detection by the enemy. (TA) While hasty positions in open areas are the least desirable, mission accomplishment may require assuming a post in an undesirable area. Under these circumstances, extreme care must be taken to utilize the terrain (ditches, depressions, and bushes) to provide maximum concealment. The utilization of camouflage nets and covers can provide additional concealment to avoid detection. There should be no limitation to ingenuity of the sniper team in selection of a hasty sniper post. Under certain circumstances it may be necessary to fire from trees, rooftops, steeples, under logs, from tunnels, in deep shadows, and from buildings, swamps, woods and an unlimited variety of open areas.

3. Position Safety. Selection of a well covered or concealed position is not a guarantee of the sniper's safety. He must remain alert to the danger of self-betrayal and must not violate the following security precautions.

a. When the situation permits, select and construct a sniper position from which to observe and shoot. The slightest movement is the only requirement for detection, therefore even during the hours of darkness caution must be exercised as the enemy may employ night vision equipment and sound travels great distances at night.

b. The sniper should not be located against a contrasting background or near prominent terrain features, these are usually under observation or used as registration points.

c. In selecting a position, consider those areas that are least likely to be occupied by the enemy.

d. The position must be located within effective range of the expected targets and must afford a clear field of fire.

e. Construct or employ alternate positions where necessary to effectively cover an area.

f. Assume at all times that the sniper position is under enemy observation. Therefore while moving into position the sniper team should take full advantage of all available cover and concealment and practical individual camouflage discipline. i. e. face and exposed skin areas

camouflaged with appropriate material. The face veil should be completely covering the face and upon moving into position the veil should cover the bolt receiver and entire length of the scope.

g. Avoid making sound.

h. Avoid unnecessary movement unless concealed from observation.

i. Avoid observing over a skyline or the top of cover or concealment which has an even outline or contrasting background.

j. Avoid using the binoculars or telescope where light may reflect from lenses.

k. Avoid moving foilage concealing the position when observing.

l. Observe around a tree from a position near the ground.

m. Stay in the shadow when observing from a sniper post within a building.

n. Careful consideration must be given to the route into or out of the post. A worn path can easily be detected. The route should be concealed and if possible a covered route acquire.

o. When possible, choose a position so that a terrain obstacle lies between it and the target and/or known or suspected enemy location.

p. While on the move and subsequently while moving into or out of position all weapons will be loaded with a round in the chamber and the weapon on safe.

4. Actions in Position. After arriving in position and conducting their hasty then detailed searches, the sniper team organizes any and all equipment in a convenient manner so it is readily accessable if needed. The sniper team continues to observe and collect any and all pertinent information for intelligence purposes. They establish their own system for observation, eating, sleeping, resting and making head calls when necessary. This is usually done in time increments of 30 to 60 minutes and worked alternately between the two snipers for the entire time they are in position, allowing one of the individuals to relax to some degree for short periods. Therefore it is possible for the snipers to remain effective for longer periods of time.

The sniper team must practice noise discipline at all times while occupying their position. Therefore arm and hand signals are widely used as a means of communicating. The following are recommended for use when noise discipline is of the utmost importance.

a. Pointing at oneself; meaning I, me, mine.

b. Pointing at partner; meaning you, your, yours.

c. Thumbs up; meaning affirmative, yes, go.

d. Thumbs down; meaning negative, no, no go.

e. Hands over eyes; meaning cannot see.

f. Pointing at eyes; meaning look, see, observe.

g. Slashing stroke across throat; meaning dead, kill.

h. Hands cupped together; meaning together.

i. Hand cupped around ear, palm facing forward; meaning listen, hear.

j. Fist; meaning stop, halt, hold up.

k. Make pumping action with arm; meaning double time.

OPPORTUNITY FOR QUESTIONS

SUMMARY

1. Reemphasize. During this period of instruction we discussed position selection and the two factors necessary to all positions (1) Provides maximum fields of observation and fire to the sniper. (2) It provides maximum concealment from enemy observation.

We then covered selection of hasty positions and that mission accomplishment might require of hasty positions and that mission accomplishment might require assuming a position in an undesireable area. All available terrain should be used to provide maximum concealment under these circumstances.

In conclusion we covered a number of safety precautions to be considered while on the move and in the process of moving into and out of position.

2. Remotivate. How well the sniper team accomplishes the mission depends, to a large degree, on their knowledge, understanding, and application of the various field techniques or skills that allows them to move, hide, observe, and detect. These skills are a measure of the sniper's ability to survive.

RANGE ESTIMATION TECHNIQUES

INTRODUCTION

1. Gain Attention. Everyone has had to estimate the distance from one point to another at some time. Usually an estimate was made either because no tool was available for exact measurements or because time did not allow such a measurement.

2. As a sniper, in order to engage a target accurately, you will be required to estimate the range to that target. However, unlike with many of your early experiences, a "ball-park guesstimate" will no longer suffice. You will have to be able to estimate ranges out to 1000 yards with 90% accuracy.

3. Purpose

a. Purpose. To acquaint the student with the various techniques of range estimation he will use in his role as a sniper.

b. Main Ideas. Describe the use of maps, the 100 meter method, the appearance of objects method, the bracketing method, the averaging method, the range card method and the use of the scope reticle in determining range.

4. Training Objectives. Upon completion of this period of instruction, the student will:

a. Determine range with the aid of a map.

b. Demonstrate the other techniques for determining range by eye.

c. Identify those factors which effect range estimation.

TRANSITION. The sniper's training must concentrate on methods which are adaptable to the sniper's equipment and which will not expose the sniper.

BODY A. MEHTODS OF RANGE ESTIMATION

1. Use of Maps. When available, maps are the most accurate aid in determining range. This is easily done by using the paper-strip method for measuring horizontal distance.

2. The 100 Meter Unit of Measure Method.

a. Techniques. To use this method, the sniper must be able to visualize a distance of 100 meters on the ground. For ranges up to 500 meters, he

determines the number of 100 meter increments between the two points. Beyond 500 meters, he selects a point midway to the targets, determine the number of 100 meter increments to the halfway point, and doubles the result.

b. Ground which slopes upward gives the illusion of greater distance, while ground sloping downward gives an illusion of shorter than actual distance.

c. Attaining Proficiency. To become proficient with this method of range estimation, the sniper must measure off several 100 meter courses on different types of terrain, and then, by walking over these courses several times, determines the average number of paces required to cover the 100m of the various terrains. He can then practice estimation by walking over unmesured terrain, counting his paces, and marking off 100m increments. Looking back over his trail, he can study the appearance of the successive increments. Conversely, he can estimate the distance to a given point, walk to it counting his paces, and thus check his accuracy.

d. Limitations. The greatest limitation to the 100m unit of measure method is that it's accuracy is directly related to how much of the terrain is visible to the observer. This is particularly important in estimating long ranges. If a target appears at a range of 100 meters or more, and the observer can only see a portion of the ground between himself and the target, the 100m unit of measure cannot be used with any degree of accuracy.

3. Appearance-of-Objects Method.

a. Techniques. This method is a means of determining range by the size and other characteristic details of some object. For example, a motorist is not interested in exact distance, but only that he has sufficient road space to pass the car in front of him safely. Suppose however, that a motorist knew that a distance of 1 kilometer (Km), an oncoming vehicle appeared to be 1 inch high, 2 inches wide, with about ½ inch between the headlights. Then, any time he saw oncoming vehicles that fit these dimensions, he would know that they were about 1 Km away. This same technique can be used by snipers to determine range. Aware of the sizes and details of personnel and equipment at known ranges, he can compare these characteristics to similar objects at unknown distances, and thus estimate the range.

b. To use the appearance-of-objects method with any degree of accuracy, the sniper must be thoroughly familiar with the characteristic details of objects as they appear at various ranges. For example, the sniper should study the appearance of a man at a range of 100 meters. He fixes the man's appearance firmly in his mind, carefully noting details of size and the characteristics of uniform and equipment. Next, he studies the same man in the kneeling position and then in the prone position. By comparing the appearance of these positions at known ranges from 100-500m, the sniper can establish a series of mental images which will help him in determining ranges on unfamiliar terrain. Practice time should also be devoted to the appearance of other familiar objects such as weapons and vehicles.

c. Limitations. Because the successful use of this method depends upon visibility, or anything which limits visibility, such as smoke, weather or darkness, will also limit the effectiveness of this method.

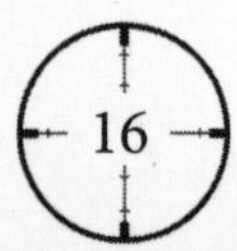

4. <u>Combination of Methods</u>. Under proper conditions, either the 100m unit of measure or the appearance-of-objects method of determining range will work, however, proper conditions rarely exist on the battlefield. Consequently, the sniper will be required to use a combination of methods. Terrain can limit the accuracy of the 100m unit of measure method and visibility can limit the appearance-of-objects method. For example, an observer may not be able to see all of the terrain out to the target, but he may see enough to get a fair idea of the distance. A slight haze may obscure many of the target details, but the observer can still make some judgment of it's size. Thus, by carefully considering the results of both methods, an experienced observer should arrive at a figure close to the true range.

5. <u>Bracketing Method</u>. By this method, the sniper assumes that the target is no more than "X" meters, but no less than "Y" meters away; he uses the average as the estimation of range.

6. <u>Averaging Method</u>. Snipers can increase the accuracy of range estimation by eye by using an average of the individual team members estimations.

7. <u>Range Card Method</u>. Information contained on prepared range cards establishes reference points from which the sniper can judge ranges rapidly and accurately. When a target appears, it's position is determined in relation to the nearest object or terrain feature drawn on the range card. This will give an approximation of the targets range. The sniper determines the difference in range between the reference point and the target, and sets his sights for the proper range, or uses the correct hold off.

8. <u>Range Estimation Formula Method</u>. This method requires the use of either binoculars or telescopic sights equipped with a mil scale. To use the formula, the sniper must know the average size of a man or any given piece of equipment and he must be able to express the height of the target in yards. The formula is:

$$\frac{\text{SIZE OF OBJ. (IN YDS) X 1000}}{\text{SIZE OF OBJ. (IN MILS)}} = \text{RANGE TO TARGET}$$

For example: A sniper, looking through his scope sees a man standing. He measures the size of the man, using the mil scale on the reticle, and he sees that the man is 4 mils high. He knows that the average man is five and a half feet tall. To convert 6 feet to yards, he divides by 3 and finds that the man is 2.0 yards tall. Using the Formula:

$$\frac{\text{SIZE OF OBJ (IN YDS) } 2.0 \times 1000}{\text{SIZE OF OBJ (IN MILS) } 4} = \frac{2000}{4} = 500 \text{ yards}$$

Once the formula is understood, the sniper needs only to be able to estimate the actual height of any target and he can determine the range to that target extremely accurately.

b. <u>Limitations</u>. While this formula can be extremely accurate, it does have several limitations.

(1) At long ranges, measurement in mils must be precise to the nearest half mil or a miss will result. For example; If a man standing appears to the $1\frac{1}{2}$ mils high, he is 1333 yds away. If he appears to be 2 mils high, he is only 1000 yds away. Careless measurement could result in a range estimation error of 333 yds in this case.

(2) This formula can be worked quickly, even if the computations are done mentally. However, as with any formula, care must be taken in working it or a totally wrong answer can result, and

(3) The formula depends entirely on the sniper's ability to estimate the actual height of a target in yards.

B. FACTORS EFFECTING RANGE ESTIMATION

1. Nature of the Target

a. An object of regular outline, such as a house, will appear closer than one of irregular outline, such as a clump of trees.

b. A target which contrasts with it's background will appear to be closer than it actually is.

c. A partially exposed target will appear more distant than it actually is.

2. Nature of Terrain. The observer's eye follows the irregularities of terrain conformation, and he will tend to overestimate distance values. In observing over smooth terrain such as sand, water, or snow, his tendency is to underestimate.

3. Light Conditions. The more clearly a target can be seen, the closer it appears. A target in full sunlight appears to be closer than the same target when viewed at dusk or dawn, through smoke, fog or rain. The position of the sun in relation to the target also affects the apparent range. When the sun is behind the viewer, the target appears to be closer. When the sun is behind the target, the target is more difficult to see, and appears to be farther away.

OPPORTUNITY FOR QUESTIONS AND COMMENTS

SUMMARY

1. Reemphasize. We have seen various ways to estimate range. Each one of them works well under the conditions for which it was devised, and when used in combination with one another, will suit any condition of visibility or terrain.

2. Remotivate. The accuracy of the shot you will fire will depend to a large extent on whether or not you have applied the rules for range estimation. Remember, if you cannot determine how far your target is away from you, you would just as well have left your rifle in the armory.

METRIC/ENGLISH
EQUIVELANTS

METRIC 1 MOA (CM)	YDS	YDS/METERS	METERS	ENGLISH 1 MOA (IN)
3	109	100	91	1
4.5	164	150	137	1.5
6	219	200	183	2
7.5	273	250	228	2.5
9	328	300	274	3
10.5	383	350	320	3.5
12	437	400	365	4
13.5	492	450	411	4.5
15	546	500	457	5
16.5	602	550	503	5.5
19	656	600	549	6
19.5	711	650	594	6.5
21	766	700	640	7
22.5	820	750	686	7.5
24	875	800	731	8
25.5	929	850	777	8.5
27	984	900	823	9
28.5	1039	950	869	9.5
30	1094	1000	914	10
31.5	1148	1050	960	10.5
33	1203	1100	1005	11

RANGE ESTIMATION TABLE FOR SIX FOOT MAN

Average Standing Man - 6 Feet Tall/2 Yards Tall

Average Sitting/Kneeling Man - 3 Feet Tall/1 Yard Tall

HEIGHT IN MILS	STANDING RANGE	SITTING/KNEELING RANGE
1	2000	1000
1.5	1333	666
2	1000	500
2.5	800	400
3	666	333
3.5	571	286
4	500	250
4.5	444	222
5	400	200
5.5	364	182
6	333	167
6.5	308	154
7	286	143
7.5	267	133
8	250	125
8.5	235	118
9	222	111
9.5	211	105
10	200	100

RANGE ESTIMATION TABLE OF MILS
FOR PERSONNEL - 6', 5'9" and 5'6"

MILS	6'-2YDS	5'9"-1.9 YDS	5'6"-1.8 YDS
1	2000	1900	1800
1-1/4	1600	1520	1440
1-1/2	1333	1266	1200
1-3/4	1143	1085	1028
2	1000	950	900
2-1/4	888	844	800
2-1/2	800	760	720
2-3/4	727	690	654
3	666	633	600
3-1/4	615	584	553
3-1/2	571	542	514
3-3/4	533	506	480
4	500	475	450
4-1/4	470	447	423
4-1/2	444	422	400
4-3/4	421	400	378
5	400	380	360
5-1/4	380	361	342
5-1/2	362	345	327
5-3/4	347	330	313
6	334	316	300
6-1/4	320	304	288
6-1/2	308	292	277
6-3/4	296	281	266
7	286	271	257
8	250	237	225
9	222	211	200
10	200	100	180

TECHNIQUES
OF
CAMOUFLAGE

INTRODUCTION

1. Gain Attention. Most uninformed people envision a sniper to be a person with a high powered rifle who either takes pot shots at people from high buildings or ties himself in a coconut tree until he's shot out of it. But to the enemy, who knows the real capabilities of a sniper, he is a very feared ghostly phantom who is never seen, and never heard until his well aimed round cracks through their formation and explodes the head of their platoon commander or radio man. A well trained sniper can greatly decrease the movement and capabilities of the most disciplined troops because of the fear of this unseen death.

2. This gives one example of how effective snipers can be and how the possibility of their presence can work on the human mind. But, marksmanship is only part of the job. If the sniper is to be a phantom to the enemy, he must know and apply the proper techniques of camouflage. He cannot be just good at camouflage. He has to be perfect if he is to come back alive.

3. Purpose

a. Purpose. The purpose of this period of instruction is to provide the student with the basic knowledge to be able to apply practically, the proper techniques of camouflage and concealment needed to remain undetected in a combat environment.

b. Main Ideas. The main ideas which will be discussed are the following:

(1) Target Indicators
(2) Types of Camouflage
(3) Geographical Areas
(4) Camouflage During Movement
(5) Tracks and Tracking

4. Training Objectives. Upon completion of this period of instruction, the student will:

a. Camouflage his uniform and himself by using traditional or expedient methods as to resemble closely the terrain through which he will move.

b. Camouflage all of his equipment to present the least chance or detection.

c. Know and understand the principles which would reveal him in combat and how to overcome them.

d. Understand the basic principles of tracking and what information can be learned from tracking.

BODY

1. Target Indicators

a. General. A target indicator is anything a sniper does or fails to do that will reveal his position to an enemy. A sniper has to know and understand these indicators and their principles if he is to keep from being located and also that he may be able to locate the enemy. Additionally, he must be able to read the terrain to use the most effective areas of concealment for movement and firing positions. Furthermore, a sniper adapts his dress to meet the types of terrain he might move through.

b. Sound. Sound can be made by movement, equipment rattling, or talking. The enemy may dismiss small noises as natural, but when someone speaks he knows for certain someone is near. Silencing gear should be done before a mission so that it makes no sound while running or walking. Moving quietly is done by slow, smooth, deliberate movements, being conscious of where you place your feet and how you push aside bush to move through it.

c. Movement. Movement in itself is an indicator. The human eye is attracted to movement. A stationary target may be impossible to locate, a slowly moving target may go undetected, but a quick or jerky movement will be seen quickly. Again, slow, deliberate movements are needed.

d. Improper Camouflage. The largest number of targets will usually be detected by improper camouflage. They are divided into three groups.

(1) Shine. Shine comes from reflective objects exposed and not toned down, such as belt buckles, watches, or glasses. The lenses of optical gear will reflect light. This can be stopped by putting a paper shade taped to the end of the scope or binoculars. Any object that reflects light should be camouflaged.

(2) Outline. The outline of items such as the body, head, rifle or other equipment must be broken up. Such outlines can be seen from great distances. Therefore, they must be broken up into features unrecognizable, or unnoticable from the rest of the background.

(3) Contrast with the Background. When using a position for concealment, a background should be chosen that will absorb the appearance of the sniper and his gear. Contrast means standing out against the background, such as a man in a dark uniform standing on a hilltop against the sky. A difference of color or shape from the background will usually be spotted. A sniper must therefore use the coloring of his background and stay in shadows as much as possible.

2. Types of Camouflage

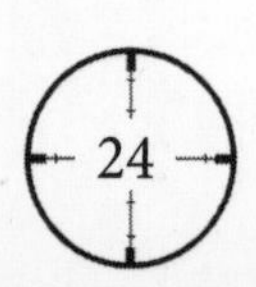

a. Stick Camouflage. In using the "grease paint", all the exposed skin should be covered, to include the hands, back of the neck, ears, and face. The parts of the face that naturally form shadows should be lightened. The predominate features that shine, should be darkened, such as the forehead, cheeks, nose and chin. The pattern and coloring that should be used is one that will blend with the natural vegetation and shadows. For jungle or woodland, dark and light green are good. White and gray should be used for snow areas, and light brown and sand coloring for deserts.

(1) Types of Patterns. The types of facial patterns can vary from irregular stripes across the face to bold splotching. The best pattern, perhaps, is a combination of both stripes and splotches. What one does not want is a wild type design and coloring that stands out from the background.

b. Camouflage Clothing.

(1) The Ghillie Suit. The guillie suit is an outstanding form of personal camouflage. It is used by both the British and Canadian Snipers to enable them to stalk close to their targets undetected. The ghillie suit is a camouflage uniform or outer smock that is covered with irregular patterns of garnish, of blending color, attached to it. It also has a small mesh netting sewn to the back of the neck and shoulders, and then draped over the head for a veil. The veil is used while in position to break up the outline of the head, hid the rifle scope, and allow movement of the hands without fear of detection. The veil when draped over the head should come down to the stomach or belt and have camouflaged garnish tied in it to break up the outline of the head and the solid features of the net. When the sniper is walking, he pushes the veil back on his head and neck so that he will have nothing obstructing his vision or hindering his movements. The veil is, however, worn down while crawling into position or while near the enemy. The ghillie suit, though good, does not make one invisible. A sniper must still take advantage of natural camouflage and concealment. Also wearing this suit, a sniper would contrast with regular troops, so it would only be worn when the sniper is operating on his own.

(2) Field Expedients. If the desired components for the construction of a ghillie suit are not on hand, a make-shift suit can be made by expedient measures. The garnish can be replaced by cloth discarded from socks, blankets, canvas sacks, or any other material that is readily available. The material is then attached to the suit in the same way. What is important is that the texture and outline of the uniform are broken. The cloth or any other equipment can be varied in color by using mud, coffee grounds, charcoal, or dye. Oil or grease should not be used because of their strong smell. Natural foliage helps greatly when attached to the artificial camouflage to blend in with the background. It can be attached to the uniform by elastic bands sewn to the uniform or by the use of large rubber bands cut from inner tubes. Care must be taken that the bands are not tight enough to restrict movement or the flow of blood. Also as foliage grows old, or the terrain changes, it must be changed.

c. Camouflaging Equipment

(1) One of the objects of primary concern for camouflage is the rifle. One has to be careful in camouflaging the rifle that the operation

is not interfered with, the sight is clear, and nothing touches the barrel. Camouflage netting can be attached to the stock, scope and sling, then garnish tied in it to break up their distinctive outline.

The M-16 and M-14 can be camouflaged in the same way ensuring that the rifle is fully operational.

(2) Optical Gear such as a spoting scope and binoculars are camouflaged in the same manner. The stand is wrapped or draped with netting and then garnish is tied into it. Make sure that the outline is broken up and the colors blend with the terrain. The binoculars are wrapped to break their distinctive form. Since the glass reflects light, a paper hood can be slipped over the objective lens on the scope or binoculars.

(3) Packs and Web Gear. Web gear can be camouflaged by dying, tying garnish to it, or attaching netting with garnish. The pack can be camouflaged by laying a piece of netting over it, tied at the top and bottom. Garnish is then tied into the net to break up the outline.

3. Geographic Areas

a. General. One type of camouflage naturally can not be used in all types of terrain and geographic areas. Before operations in an area, a sniper should study the terrain, vegetation and lay of the land to determine the best possible type of personal camouflage.

(1) Snow. In areas with heavy snow or in wooded areas with trees covered with snow, a full white camouflage suit is worn. With snow on the ground and the trees are not covered, white trousers and green-brown tops are worn. A hood or veil in snow areas is very effective. Firing positions can be made almost totally invisible if made with care. In snow regions, visibility during a bright night is as good as in the day.

(2) Desert. In sandy and desert areas, texture camouflage is normally not necessary, but full use of the terrain must be made to remain unnoticed. The hands and face should be blended into a solid tone using the proper camouflage stick, and a hood should be worn.

(3) Jungle. In jungle areas, foliage, artificial camouflage, and camouflage stick are applied in a contrasting pattern with the texture relative to the terrain. The vegetation is usually very thick so more dependence can be made on using the natural foliage for concealment.

4. Camouflage During Movement.

a. Camouflage Consciousness. The sniper must be camouflage conscious from the time he departs on a mission until the time he returns. He must constantly observe the terrain and vegetation change. He should utilize shadows caused by vegetation, terrain features, and cultural features to remain undetected. He must master the techniques of hiding, blending, and deceiving.

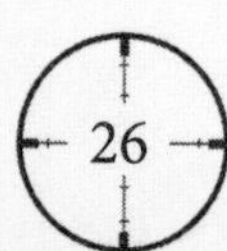

(1) Hiding. Hiding is completely concealing yourself against observation by laying yourself in very thick vegetation, under leaves, or however else is necessary to keep from being seen.

(2) Blending. Blending is what is used to the greatest extent in camouflage, since it is not always possible to completely camouflage in such a way as to be indistinguishable from the surrounding area. A sniper must remember that his camouflage need be so near perfect that he should fail to be recognized through optical gear as well as with the human eye. He must be able to be looked at directly and not be seen. This takes much practice and experience.

(3) Deceiving. In deceiving, the enemy is tricked into false conclusion regarding the sniper's location, intentions, or movement. By planting objects such as ammo cans, food cartons, or something to intrique the enemy, that he may be decoyed into the open where he can be brought under fire. Mannequins can be used to lure the enemy sniper into firing, thereby revealing his position.

5. Tracks and Tracking.

a. General. Once a sniper has learned camouflage and concealment to perfection, he must go one step further. This is the aspect of him leaving no trace of his presence, activities or passage in or through an area. This is an art in itself, and is closely related to tracking, which can tell you in detail about the enemy around you.

(1) Enemy Trackers or Scouts. It is said that the greatest danger to a sniper is not the regular enemy soldier, but in fact, is hidden booby traps, and the enemy scout who can hunt the sniper on his own terms. If an enemy patrol comes across unfamiliar tracks in it's area of operation, it may be possible for them to obtain local trackers. If it is a man's livelihood to live by hunting, he will usually be very adept at tracking. What a professional can read from a trail is truly phenominal. Depending on the terrain, he will be able to determine the exact age of the trail, the number of persons in the party, if they are carrying heavy loads, how well trained they are by how well they move, their nationality, by their habits and boot soles, how fast they are moving and approximately where they are at the moment. If a tracker determines a fresh trail to be a party of four, who, but recon and sniper teams move in such small groups behind enemy lines? The enemy will go to almost any extreme to capture or kill them.

(2) Hiding Personal Signs. A modern professional tracker who makes his living of trailing lost children, hunters, or escaped convicts, was once asked, "Can a man hide his own trail well enough that a tracker cannot follow him?" The professional tracker answered, "NO, there is no way to hide a trail from a true tracker." The chances in combat of being pitted against a "real tracker" are rare, but all it takes is one time. This is to emphasize the importance of leaving no signs at all for the enemy scout to read. This is done by paying particular attention to where and how you walk, being sure not to walk in loose dirt or mud if it can be avoided, and not scuffing the feet. Walking on leaves, grass, rocks, etc. can help hide tracks. Trails are also made by broken vegetation

such as weeds, limbs, scrape marks on bushes, and limbs that have been bent in a certain direction. When moving through thick brush, gently move the brush forward, slip through it, then set it back to it's normal position. Mud or dirt particles left on rocks or exposed tree roots are a sign of one's presence. Even broken spider webs up to the level of a man's height show movement. In the process of hiding his trail, a sniper must remember to leave no debris such as paper, C-Ration cans, spilled food, etc. behind him. Empty C-Ration cans, can either be carried out, or smashed, buried, and camouflaged. Along this same line, a hole should be dug for excrement, then camouflaged. The smell of urine on grass or bushes lasts for many days in a hot humid environment, so a hole should be dug for this also. One last object of importance, the fired casings from the sniper rifle must always be brought back, for they are a sure sign of a sniper's presence.

(3) Reading Tracks and Signs. To be proficient at tracking takes many years of experience, but a knowledgeable sniper can gain much information from signs left by the enemy. For instance, he can tell roughly the amount of enemy movement through a given area, what size units they move in, and what areas they frequent the most. If an area is found where the enemy slept, it may be possible to determine the size of the unit, how well disciplined they are, by the security that was kept, and their overall formation. It can be fairly certain that the enemy is well fed if pieces of discarded food or C-Ration cans with uneaten food in them are found. The opposite will also be true for an enemy with little food. Imprints in the dirt or grass can reveal the presence of crew-served weapons, such as machine guns or mortars. Also, prints of ammo cans, supplies, radio gear may possibly be seen. The enemies habits may come to light by studying tracks so that he may be engaged at a specific time and place.

OPPORTUNITY FOR QUESTIONS AND COMMENTS

SUMMARY

1. Reemphasize. During this period of instruction, we have discussed target indicators, types of camouflage, geographical areas, and tracks. Initially, we covered what a target indicator is, and that sound, movement, and improper camouflage make up indicators. Care must be taken that shine, outline, and contrast with the background are eliminated.

We learned the different types of camouflage, such as grease paint, and how totone down the skin with it, the ghillie suit, used as camouflaged clothing, and field expedient measures for camouflaging clothing and equipment. In the section on geographical areas, we learned the different types of camouflage used in the various climate regions.

The fourth area covered was concealment during movement and how to use terrain features. We learned the difference in hiding, blending, and deceiving, and how to use each. We learned of the danger of enemy scouts or trackers, and the importance of leaving no indication of one's presence while on a mission. Lastly, we covered what a sniper can learn from enemy tracks if he is observant enough to see them and takes the time to learn their meaning.

2. Remotivate. The job of a sniper is not for a person who just wants the prestige of being called a sniper. It is a very dangerous position even if the sniper is well trained and highly motivated. Expertise at camouflage to remain unnoticed takes painstaking care, and thoroughness which the wrong type of individual would not take time to do. If you are to be successful at camouflage and concealment, it takes a double portion in carefulness on your part, if you are to come back alive.

3. Concealment

a. Concealed Approaches. It is essential that the natural appearance of the ground remains unaltered, and that any camouflage done is of the highest order. The sniper must also remember that though cover from view is cover from aimed fire, all concealment will be wasted if the sniper is observed as he enters the hide. It follows, therefore, that concealed approaches to the hide are an important consideration, and movement around it must be kept to a minimum. Efforts must be made to restrict entry to and exit from the hide prior to darkness. Track discipline must be rigidly enforced.

b. Screens. Any light shining from the rear of a hide through the front loophole may give the position away. It is necessary, therefore, to put a screen over the entrance to the hide, and also one over the loophole itself. The two screens must never be raised at the same time. Snipers must remember to lower the entrance screen as soon as they are in the hide and to lower the loophole screen before leaving it. These precautions will prevent light from shining directly through both openings.

c. Loopholes. Loopholes must be camouflaged using foliage or other material which blends with or is natural to the surroundings. Logically, anything not in keeping with the surroundings will be a source of suspicion to the enemy and hence a source of danger to the sniper.

d. Urban Areas. In urban areas, a secure and quiet approach with the minimum number of obstacles such as crumbling wall and barking dogs is required. When necessary, a diversion in the form of a vehicle or house search can be set up to allow the sniper the use of the cover of a vehicle to approach the area unseen and occupy the hide.

4. The Use Of Buildings And Hides As Fire Positions.

a. Disadvantages. Buildings can often offer good opportunities as sniping posts under static conditions. they suffer, however, the great disadvantage that they may be the object of attention from the enemy's heavier weapons. Isolated houses will probably be singled out even if a sniper using it has not been detected.

b. Preparation. Houses should be prepared for use in much the same way as other hides, similar precautions towards concealment being taken; loopholes being constructed and fire positions made.

c. Outward Appearance. Special care must be taken not to alter the outward appearance of the house by opening windows or doors that were found closed, or by drawing back curtains.

d. Free Positions. The actual fire positions must be well back in the shadow of the room against which the sniper might be silhouetted, must be screened.

e. Loopholes. Loopholes may be holes in windows, shutters of the roof, preferable those that have been made by shells of other projectiles. If such loopholes have be be picked out of a wall, they must be made to look like war damage.

f. Observation Rest. Some form of rest for the firer and observer will have to be constructed in order to obtain the most accurate results. Furniture from the house, old mattresses, bedspreads and the like will serve the purpose admirably; if none of this material is available, sand-bags may have to be used.

5. Firing From Hides.

a. Fire Discipline. Fire from a hide must be discreet and only undertaken at specific targets. Haphazard harassing fire will quickly lead to the enemy locating the hide and directing fire to it.

b. Muzzle Flash. At dusk and dawn, the flash from a shot can usually be clearly seen and cars must be taken not to disclose the position of the hide when firing under such circumstances.

c. Rifle Smoke. On frosty mornings and damp days, there is a great danger of smoke from the rifle giving the position away. On such occasions, the sniper must keep as far back in the hide as possible.

d. Dust. When the surroundings are dry and dusty, the sniper must be careful not to cause too much dust to rise. It may be necessary to dampen the surroundings of the loophole and the hide when there is a danger of rising dust.

6. Types of Hides. A hide can take many forms. The type of operation or the battle situation coupled with the task the snipers are given, plus the time available, the terrain and above all, the ingenuity and inventiveness of the snipers, will decide how basic or elaborate the hide can or must be. In all situations, the type of hide will differ, but the net result is the same, the sniper can observe and fire without being detected.

a. Belly Hide. This type hide is best used in mobile situations or when the sniper doesn't plan to be in position for any extended period of time. Some of the advantages and disadvantages are:

(1) It is simple and can be quickly built.
(2) Good when the sniper is expected to be mobile, because many can be made.

Disadvantages

(1) It is uncomfortable and cannot be occupied for long periods of time.

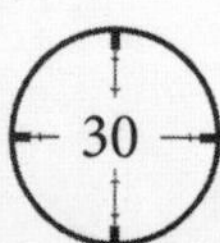

b. Enlarged Fire Trench Hide. This type hide is nothing more than an enlarged fighting hole with advantages being:

(1) Able to maintain a low silhouette,
(2) Simple to construct,
(3) Can be occupied by both sniper and observer,
(4) It can be occupied for longer periods of time with some degree of comfort.

Disadvantages

(1) It is not easily entered or exited from.
(2) There is no overhead cover when in firing position.

c. Semi-Permanent Hide. This type hide resembles a fortified bunker and should always be used if time circumstances permit. The advantages are:

(1) Can be occupied for long periods of time,
(2) Gives protection from fire and shrapnel,
(3) Enables movement for fire and observation,
(4) Provides some comfort.

Disadvantages

(1) Takes time to construct
(2) Equipment such as picks, shovels, axes, etc. are needed for construction.

d. Shell Holes. Building a hide in a shell hole saves a lot of digging, but needs plenty of wood and rope to secure the sides. Drainage is the main disadvantage of occupying a shell hole as a hide.

e. Tree Hides. In selecting trees for hides, use trees that have a good deep root such as oak, chestnut, hickory. During heavy winds, these trees tend to remain steady better than a pine which has surface roots and sways quite a bit in a breeze. A large tree should be used that is back from the wood line. This may limit your field of view, but it will better cover you from view.

OPPORTUNITY FOR QUESTIONS AND COMMENTS

SUMMARY

1. Reemphasize. During this period of instruction, we covered the complete construction of hides and locating the best area of this construction. It should be stressed that the sniper should use his own imagination and initiative while constructing his hide. In conclusion, we discussed the various types of hides, advantages and disadvantages and in what situation a particular hide could be best used.

2. Remotivate. The type of hide you build will depend on a great many things. Time, Terrain, Type of Operation, Enemy Situation, and Weapons. So always construct a good defensive hide. It will keep you effective and keep you alive.

INTRODUCTION OF NIGHT VISION DEVICES

The objective of this lesson was to enable you to engage targets during hours of limited visibility with the night vision device by proper mounting, operation, and maintenance of this piece of equipment, and explain the capabilities and limitations of each.

1. In order to effectively engage targets during hours of limited visibility, you must know proper procedure for mounting, operation, and operator maintenance of the NVD.

2. The adaptor bracket must be aligned with the mounting groove on the left side of the receiver and tightened securely with an allen wrench before the sight assembly can be installed, by rotating the lock knobs counterclockwise until they stop on the pins on the assembly. You then slide the boresight onto the guide rail until it is positioned against the pinstop. The last step in mounting is to tighten the lock knob by turning clockwise.

3. The first point to remember about operation of the NVD is the mercury battery is irritable to the eyes and other mucus membranes. Even though the image intensifier will turn off automatically to protect the observer, the image intensifier should <u>never</u> be pointed at the sun on or off. The NVD will always be inspected before use. The operational sequence is definite as should always be followed.

4. The NVD will operate from -65°F to; -125°F and the lenses will have to be cleaned frequently in sandy or dusty areas and will also operate in wet or humid areas.

5. Operator maintenance can be performed with the tools and equipment provided with the exception of a screwdriver and should be followed step-by-step. Above all, <u>never</u> use lubricating materials on the NVD.

6. To complicate training, the two scopes in the system (AN/PVS-2 and AN/PVS-4) have some important differences:

a. They require different zero procedures.

b. They have totally different reticles.

c. They have different mounting and dismounting procedures.

d. Windage and elevation adjustments on the AN/PVS-2 are made in the direction of the error, while adjustments on the AN/PVS-4 are made in the direction of the desired point of impact.

7. At 25 meters, the point of impact for the AN/PVS-2 is 1.2 cm high and 3.8 cm right. The AN/PVS-4 is 1.4 cm high.

AN/PVS-4

Horizontal line from left
point of origin
20 feet at ranges shown.
Range is in hundreds of meters.

Vertical lines above or
below horizontal line
represent 6 feet at ranges
shown. Range is in
hundreds of meters.

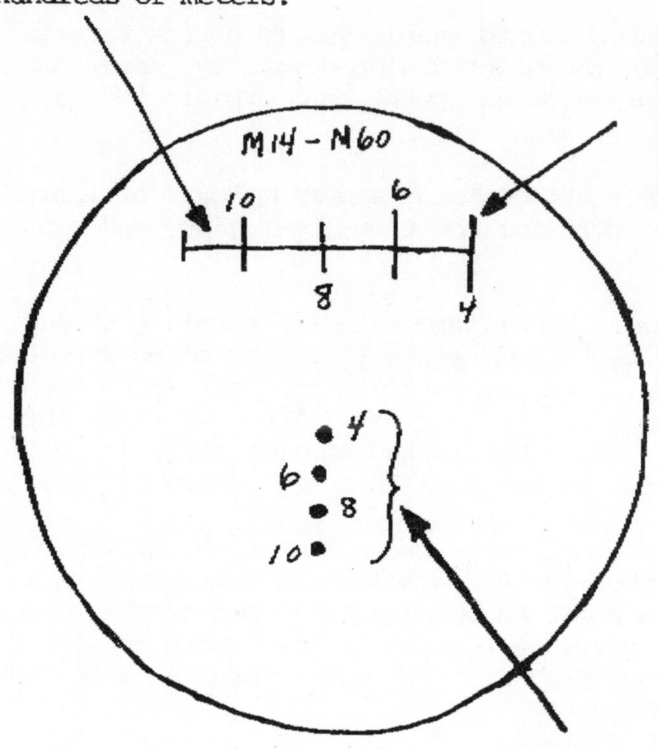

M14-M60 Aiming points.
Range is in hundreds of
meters.

Use center of two horizontal
lines for 0-250 meters.

Example:

A) Distance to tank is 800 m
B) Distance to 6' man is 600 m
C) Distance to 6' man is 200 m

Reference: TM 11-5855-211-10

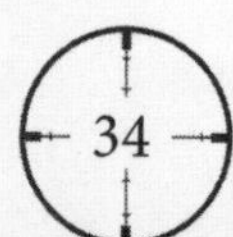

AN/PVS-2

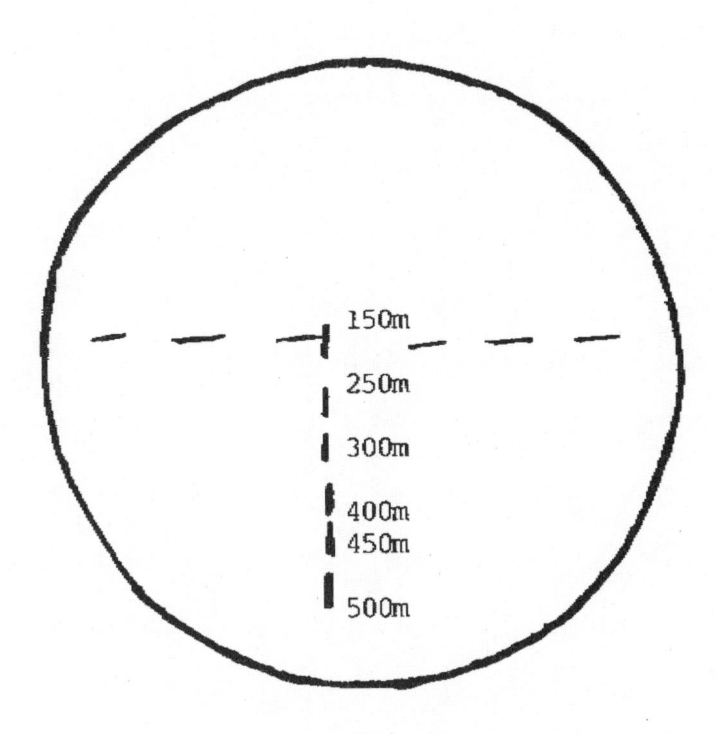

BLACK LINE RETICLE PATTERN

Through experience and test firing (zeroing), it has been determined that the placement of the reticle index marks produce the above noted range zeroing reference points.

Using these aiming points in the center of mass of a target will enable the sniper to obtain a first round hit.

Reference: TC 23-14

NAVAL SPECIAL WARFARE

SCOUT/SNIPER SCHOOL

DATE________________

LITTON MODEL M-845 NIGHT VISION WEAPONS SIGHT

1. LESSON PURPOSE. In order the SEAL sniper to engage targets during hours of limited visibility, he must use a night vision device. This period of instruction will provide you with the knowledge necessary to mount, put into operation, zero, and maintain the M-845 night vision device.

2. OBJECTIVES:

a. Lesson Objective: To enable the student to mount, put into operation, and zero the M-845 night vision device as stated in class.

b. Training Objectives.

1. Mount the M-845.

2. Place the NVD into operation.

3. Zero the M-845 NVD.

GENERAL:

The Litton Model M-845 is a compact, lightweight, battery powered night vision weapons sight for small arms use in low light conditions.

The weapons sight is effective at both short and intermediate ranges in a variety of environments and conditions. Employing a state-of-the-art 18MM 2nd generation microchannel plate image intensifier tube (light amplifier), the M-845 affords small arms users a completely "passive" night sighting and aiming capability.

The Litton M-845 is specifically designed for use with small bore weapons, (7.62MM and 5.56MM calibers). The litton night vision weapons sight is the smallest and lightest device of its type currently used by Special Warfare.

The rifle sight is furnished with a waterproof, flip-up lend cover that incorporates a daylight filter. This feature permits zeroing of the weapon during the hours of daylight.

1. NORMAL COMPLIMENT.

a. Model M-845 night vision weapons sight.

b. Lens cap/daylight filter.

c. Lens cleaning tissue.

e. Instruction and maintenance manual.

f. 2 type E-132 mercury batteries.

g. Thermo-formed ABS plastic transit and storage case.

h. Mounting kit with all tools and hardware for mounting the M-845.

2. CONTROLS AND ADJUSTMENTS.

a. ON/OFF rotary switch.

b. Diopter focus ring.

c. Elevation adjustment knob.

d. Windage adjustment knob.

3. SYSTEM.

a. Magnification..................1.55 x

3. ELECTRICAL.

a. Battery life..................Approx. 40 hours (70 degrees).

b. Low battery indicator.........Red L.E.D.

c. Battery type (one each)........Mercury - 2.8 VDC Type 132.
Lithium - 3.0 VDC Type 440S-BT.

4. MECHANICAL.

a. Length........................9.7 in.

b. Height........................3.2 in.

c. Width.........................2.6 in.

d. Weight w/battery..............2.2 lbs.

NOTE:

a. The litton M-845, at longer ranges (200 yards and beyond) the red dot will obscure a man size target.

PRELIMINARY DATA SHEET

4X Night Binocular, Model M975, M976

System Performance

Magnification:	4.2 x
Field of View:	8.5 degrees
Focus Range:	30 m to infinity
Limiting Resolution:	3.0 lp/mR

*Detection Range (m)	Full Moon		Starlight/Overcast	
	Gen II	Gen III	Gen II	Gen III
Man	1140	1500	380	640
Tank	2850	4000	950	1600

Objective Lens

Focal Length:	116 mm
T-Number:	2.0

Image Intensifier	M975	M976
Type:	Gen II plus, non-inverting	Gen III, non-inverting
Gain:	18,000 - 25,000	20,000 - 35,000
Resolution (min):	28 lp/mm	28 lp/mm

* Detection ranges are calculated from laboratory tests under controlled light levels. Actual field performance may vary depending on atmospheric conditions.

b. The M-845 does not have the light gathering capabilities of the PVS-4 night vision device.

c. The M-845 also does not have a range finding capability as compared to the PVS-4 night vision device.

5. MOUNTING PROCEDURES.

a. Attach the adapter to the base of the M-845 using the #10-24 x .312 L. socket head cap screws.

NOTE.

The use of a thread locking compound such as 222 locktight is recommended.

b. Place the adapter on the night sight into the groove in the carrying handle.

c. Align the tapped hole in the adapter with the hole in the handle; then insert and tighten the level screw assembly.

6. ZEROING PROCEDURES.

a. Place a target at 25 meters.

b. Support the weapon in a stable firing position.

c. Turn the scope on.

d. Fire a few rounds to seat the night sight on the weapon, retightening if necessary.

e. Place the red dot over the center mass of the target and fire a three round group.

f. Adjust the elevation and windage gears as required. Each click on the elevation or windage will move the impact of the round 0.55 in at 25 meters.

g. Repeat steps e and f until the center of the impact group is at the center of the target or until the weapon is shooting point of aim point of impact.

h. Place a target at 100 meters and repeat steps e and f (at 100 meters when the weapon is zeroed at 25 meters, the point of aim point of impact will be 5 cm high).

NOTE:

Each click of the elevation and windage will move the bullet impact a greater distance at longer ranges. (1.2 inches/click per 100 meters).

Eyepiece Lens

Exit Pupil:	8 mm
Eye Relief Distance:	27 mm
Interpupillary Distance:	55 to 71 mm
Diopter Range:	-6 to +2

Mechanical

Weight:	1.1 Kg with AA batteries
Length x Max Width:	240 mm x 130 mm

Power Source

	Battery Type	Operational Life @ 20°C
	2 x AA (1.5 VDC)	60 Hours
or	1 x BA-1567/u	30 Hours
or	1 x BA-5567/u	30 Hours

Other Features

* Daylight Training Filter incorporated in objective lens cap
* Fully interchangeable with M972, M973 Night Vision Goggle
* Immersion tested 1 m water for ½ hour
* NBC mask compatible

BELOW: LITTON M845 MK II ELECTRON DEVICE
RIGHT: LITTON M845 MK II AND ACCESSORIES
BOTTOM: LITTON M975 ELECTRON DEVICE

NAVAL SPECIAL WARFARE

SCOUT SNIPER SCHOOL

DATE______________________

SNIPER EQUIPMENT

1. Rifle, REMMINGTON MODEL 700

The sniper rifle used by Naval Special Warfare is a bolt action, 7.62mm rifle with a stainless steel barrel for improved accuracy. It weighs anywhere from 9½ lbs to 12 lbs depending on the type stock used. The stocks vary in weight from 1 lb. 10 oz. to 3 lbs plus, and are constructed from fiberglass. The rifle is fitted with a topmounted telescope base, to which the sniper scope can be readily attached without special tools.

a. Safety lever. Is located at the rear of the receiver, behind the bolt handle.

b. When pulled to the rear, the weapon is on safe.

c. Bolt stop. The bolt stop release is located inside the trigger guard just forward of the trigger. When depressed, it allows the bolt to be removed from the rifle.

d. Floor plate latch. Is forward of the trigger guard and is opened by pressing the serrated detent on the forward edge of the trigger guard.

e. Tabulated data.

Caliber......................7.62mm NATO

Length.......................44 inches

Weight.......................9½ lbs to 12 lbs

Barrel length................24 inches

Twist, right hand...........6

Lands and groves............1 turn in 12 inches

Trigger weight..............3 to 5 lbs.

Torque.......................65 inch/pounds

Magazine capacity...........5 rounds

Max effective range.........1000 yards

2. TELESCOPIC SIGHT

GENERAL. A telescopic sight is an instrument which facilitates accurate aiming by use of precision gound lenses and crosshairs in a metal body.

1. The optical system. Is composed of a series of glass lenses which transmit and magnify the image of the target to the sniper.

2. MAGNIFICATION (Resolving power.) The average unaided eye can distinguis 1-inch detail at 100 yards. Magnification, combined with good optics design, permits resolution of this 1-inch divided by the magnification. Thus, 1/10 inch detail can be seen at 100 yards with a 10x scope.

3. LENS COATING. The Leupole & Stevens Ultra M-1 lens surfaces are coated with a high efficiency, low reflection film. This coating increases the light gathering capability to approximate 91% of the available light. With uncoated lenses, a 45% of the available light is lost in the scope.

4. FIELD OF VIEW. Field of view is the diameter of the picture seen through a scope, and it is usually in "FEET AT HUNDREDS OF YARDS."

a. ULTRA 10x-M1 field of view @ 100 meters = 3.5 meters.

5. TABULATED DATA

WEIGHT....................1,4 lbs

LENGTH....................13 1/8 inches

MAGNIFICATION............10x

EYE RELIEF...............3 inches fixed

ADJUSTMENTS (E)..........+ or - 45
(W)..........+ or - 22.5

ELEVATION AND WINDAGE....1/4 minute. 90 minutes of elevation and windage

MAIN ELEVATION...........1/4 minute resolution type adjustment

WINDAGE..................1/4 minute resolution type adjustment

RETICLE..................Mil dot duplex for range estimation and calculat leads on moving targets. (3/4 or 1/4 mil dot.)

6. FOCUS CHECK. The teescope should be focused to the individual's eye. To check the focus, point the scope at a distance scene or the sky and drape white hankerchief over the objective end. Look at a distance scene with unaided eye for several seconds and quickly glance into the eyepiece of the scope. If properly focused, the reticule should appear instantly, distinct and sharp. If not the case, the eyepiece requires focusing.

7. FOCUS OF THE EYEPIECE. To focus the eyepiece, adjust the side focus knot until the correct focus is achieved.

8. PARALLAX. Parallax is defined as the apparent movement of an object as seen from two different points not on line with the object. Observe a target at a range of 300 yards. While looking through the scope, move the head vertically and horizontally in small increments. The reticle should not appear to change position on the target. If it does, parallax is present and the objective lens must be focused.

9. PARALLAX. To focus the objective lens, adjust the side focus knob until the scope is free of parallax.

10. EYE RELIEF. When issued the telescope should be set all the way forward in the scope mounting rings. This setting will provide the needed 2 to 3 inch eye relief for almost all shooters. It is possible, however, to move the sight slightly to achieve proper eye relief by unlocking the keeper ring to the rear of the scope and adjusting the rear of the scope to the desired eye relief.

11. THE RETICLE. The duplex reticle in the telescope provides the sniper with a range-finding capability. To determine range, the following formula is used:

$$\frac{\text{Heighth of target (in meters or yards) x 1000}}{\text{Heigth of target (in mils)}} = \text{Range}$$

The dots on the fine crosshairs are 1 mil apart with a total of 5 dots from the center to the thick post in each direction.

12. ELEVATION AND WINDAGE. The 10x M1 model has approximately 90 minutes of elevation adjustments, and 15 minutes of windage, it features friction type adjustments. For example, almost twice as much needed for the 7.62 mm 173 gr. national match cartridge to reach 1000 yards. The adjustments have 1/4 minutes clicks with both audible and tactical feedback.

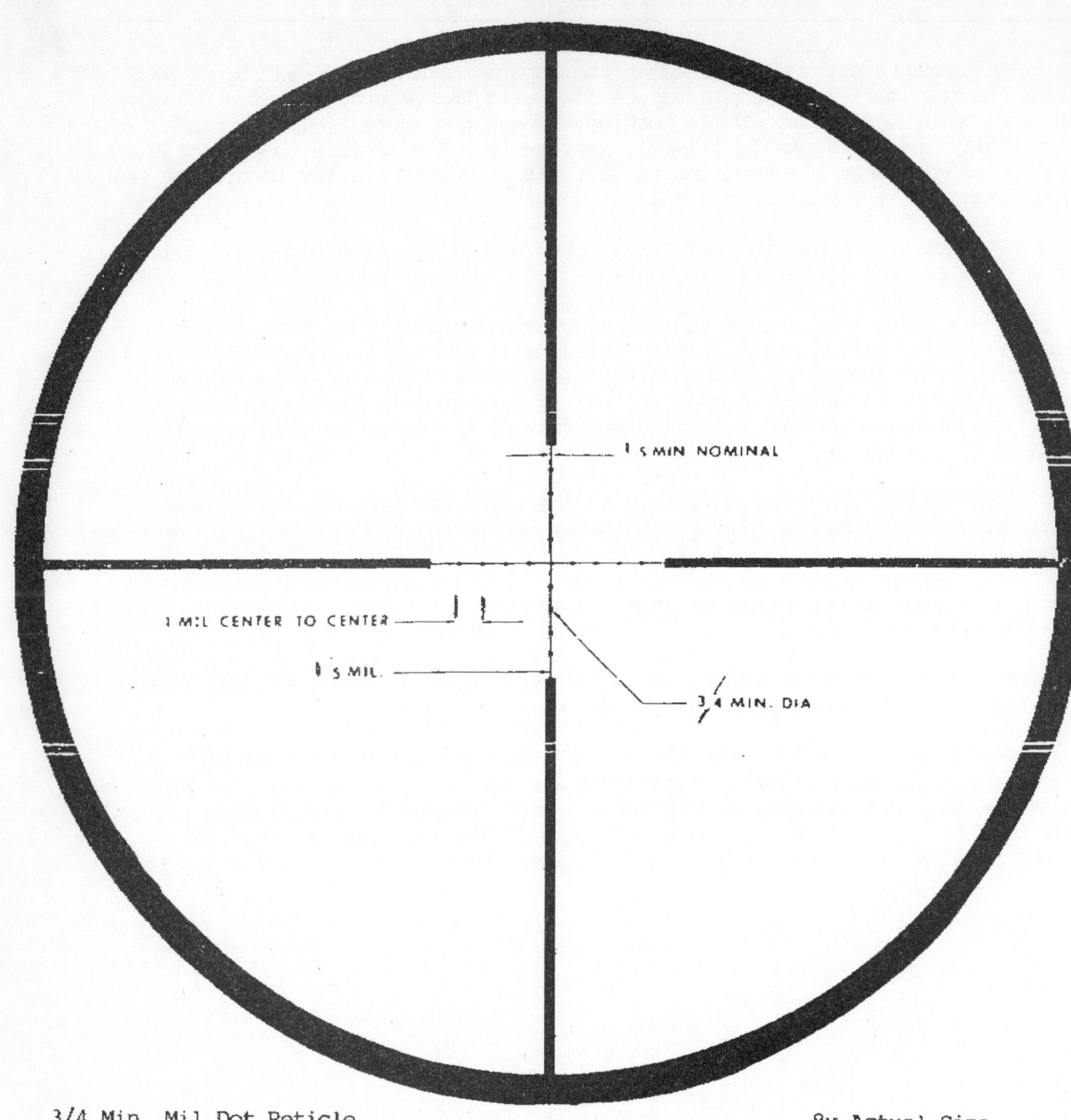

3/4 Min. Mil Dot Reticle
Fig. #3

8x Actual Size

The Mil Dot Reticle is a duplex style reticle having thick outer sections and thin center sections. Superimposed on the thin center section of the reticles is a series of dots, (4 each side of the center and 4 above and below the center) that are spaced 1 milliradian apart, and 1 milliradian from both the center and the start of the thick section of the reticle. This spacing allows the user to make very accurate estimates of target range, assuming there is an object of known size (estimate) in the field of view. For example, a human target could be assumed to be about 6' tall, which equals 1.83 meters, or at 500 meters, 3.65 dots high (nominally, about 3.5 dots high). Another example would be a 1 meter target at 1000 meters range would be the height between two dots or the width between two dots. Basically, given a good estimation of the objects size, it is possible to fairly accurately determine the target range using the mil dot system.

NAVAL SPECIAL WARFARE

SCOUT SNIPER SCHOOL

DATE________________________

OBSERVATION AIDS

The sniper's success is detecting targets, or the signs of enemy presence, is dependent on his powers of observation. To increase his ability to observe well, he is aided by the use of the telescope, binoculars, and starlight scope.

1. OBSERVATION SCOPE. The observation telescope is a prismatic optical instrument of 20-power magnification. The lenses are coated with magnesium flouride for high light gathering capability. The scope should be carried by the sniper team when it is justified by their mission. The high magnification of the telescope makes observation and target detection possible when conditions would otherwise prevent it. Camouflaged targets and those in deep shadows can be located, troop movements can be distinguished at great distances, and selected targets can be identified.

2. OPERATION. An eyepiece cap cover cap and objective lens cover are used to protect the optics when the telescope is not in use. Care must be taken to prevent cross-threading of the fine threads.

3. FOCUS. The eyepiece focusing sleeve is turned clockwise or counter-clockwise until the image can be clearly seen by the operator. On other models the focus is adjusted by the focus knob at the bottom of the telescope.

4. TRIPOD. The height adjusting collar is a desired height for the telescope. The collar is held in position by tighting the clamping screw.

The shaft rotation locking thumb screw clamps the tripod shaft at any desired azimuth.

The elevating thumbscrew is used to adjust the of the tripod, to increase, or decrease the angle of elevation of the telescope.

The tripod legs can be held in an adjusted position by tightening the screw nut at the upper end of each leg.

4. SETTING UP THE SCOPE AND TRIPOD. Spread the legs and place the tripod on a level position on the ground.

Attach the scope to the tripod by screwing the attachment screw on the tripod, clockwise until tight. To take the scope off the tripod, unscrew the attachment screw counterclockwise.

1. **BINOCULARS.** Each sniper team will be equipped with binoculars to aid in observing the enemy and in searching for and selecting targets. The focusing adjustments are on each eyepiece. The left monocle has a mil scale etched into it.

2. **METHODS OF HOLDING THE BINOCULARS.** The binoculars should be held lightly resting on and supported by the heels of the hands. The thumbs block out light that would enter between the eye and the eyepiece. The eyepieces are held lightly to the eye to avoid transmitting body movement. Whenever possible, a stationary rest should support the elbows.

An alternate method for holding the binoculars is to move the hands forward, cupping them around the sides of the objective lenses. This keeps light from reflecting off the lenses, which would reveal the sniper's position.

3. **ADJUSTMENTS.** The interpupillary distance is the distance between the eyes. The monocles are hinged together so that they can be adjusted to meet this distance. The hinge is adjusted until the field of vision ceases to be two overlapping circles and appears as single, sharply defined circle. The setting of the hinge scale should be recorded for future use.

Each individual and each eye of that individual requires different focus settings. Adjust the focus for each eye in the following manner:

a. With both eyes open, look through the glasses at a distance object.

b. Place one hand over the objective lens of the right monocle and turn the focusing ring of the left monocle until the object is sharply defined.

c. Uncover the right monocle and cover the left one.

d. Rotate the focusing ring of the right monocle until the object is sharply defined.

e. Uncover the left monocle. The object should then be clear to both eyes.

f. Read the diopter scale on each focusing ring and record for future reference.

(1) **RETICLE.** The mil scale that is etched into the left lens of the binoculars is called the reticle pattern and is used in adjusting artillery fire and determining range to a target.

(2) Determining range with the binos is done similar to the telescope sight reticle. The height of the target is measured in mils. This is then divided into the height of the target in yards (or meters), times 1000, to give the range to the target in yards (or meters). Care should be taken to measure the target to the nearest 1/4 yard.

1. ADDITIONAL USES FOR BINOCULARS. In addition to observing and adjusting fire and range estimation, binoculars may be used to:

a. Identify troops, equipment, vehicles, etc.

b. Observe enemy movement or positions

c. Make visual reconnaissance

d. Locate targets

e. Study terrain

f. Select routes and positions

g. Improve night vision

h. Improve vision in periods of reduced vision.

2. EYE FATIGUE. Prolonged use of the binoculars or telescope will cause eye fatigue, reducing the effectiveness of observation. Periods of observation with optical devices should be limited to 30 minutes by a minimum of 15 minutes of rest.

AMMUNITION

Match quality ammuntion will normaly be issued because of its greater accuracy and reduced sensity to the wind. However, if match ammuntion is not available, or the siduation dictates grade of ammunition may have to be used. In ammunition other than match, accuracy and point of aim point of impact may vary noticeably. Among different lots, an espectal accurate lot can be identified through use, and it should be used as long as it is available.

1. DATA MATCH AMMUNITION M118.

173-grain boattail bullet

Velocity..........................2,550 feet per second.

Accuracy Specification............3.5 feet mean radius at 600 yards.(1 minute of angle).

Caliber...........................Match is stamped on the head, along with the year of production and the intitials of the arsenal which produced it(e.g. L.C. ID.Lake city).

2. DATA BALL AMMUNITION M80.

147 grain bullet

Velocity..........................2,750 feet per second.

Caliber...........................7.62mm M80 and M80E1.

Identifiers.......................The year of production and the arsenal's initials are stamped on the head.

NOTE: M80E1 is the most accurate of the ball ammunition.
Because MATCH ammunition is heaver and slower than the other types, it is safe to assume that all other types of ammunition will strike higher on the target.

LAKE CITY M118 MATCH GRADE AMMUNITION WILL BE ISSUED TO THE SNIPER AND SHOULD BE FIRED AT ALL TIMES, WHEN AVAILIBLE.

TECHNIQUES OF OBSERVATION

INTRODUCTION

1. Gain Attention. The Special Warfare sniper's mission requires him to support combat operations by delivering precision fire from concealed positions to selected targets. The term "selected targets" correctly implies that the sniper is more concerned with the significance of his targets than with the number of them. In his process of observation, he will not shoot the first one available, but will index the location and identification of all the targets he can observe.

2. The sniper is expected to perform several missions other than sniping. One of the more important is observation of the enemy and his activities.

3. Purpose

a. Purpose. The purpose of this period of instruction is to provide student with the knowledge, procedures and techniques applicable to both day and night time observation.

b. Main Ideas. The main ideas which will be discussed are the following:

(1) Observation Capabilities and Limitations
(2) Observation Procedures

4. Training Objectives. At the conclusion of this period of instruction, the student, without the aid of references, will be able to:

a. Describe the limitations of observation and the steps to be taken to overcome those limitations.

b. Describe the use of the telescope, 10X Ultra, and the starlight scope as an observation aid.

c. Describe the procedures used to observe and maintain observation of a specific area or target.

TRANSITION. Observation is the keynote to a sniper's success. He must be fully aware of the human capabilities and limitations for productive observation in waning light and in darkness and of his aids, which can enhance his visual powers under those conditions.

BODY

1. Capabilities and Limitations.

a. Night Vision. Night runs the gamut from absolute, darkness to bright moonlight. No matter how bright the night may appear to be, however, it will never permit the human eye to function with daylight precision. For maximum effectiveness, the sniper must apply the proven principles of night vision.

(1) Darkness Adaptation. It takes the eye about 30 minutes to regulate itself to a marked lowering of illumination. During that time, the pupils are expanding and the eyes are not reliable. In instances when the sniper is to depart on a mission during darkness, it is recommended that he wear red glasses while in light areas prior to his departure.

(2) Off-Center vs. Direct Vision. (TA #1) Off-center vision is the technique of focusing attention on an object without looking directly at it. An object under direct gaze in dim light will blur and appear to change shape, fade, and reappear in still another form. If the eyes are focused at different points around the object and about 6 to 10 degrees away from it, side vision will provide a true picture of the object.

(3) Scanning. Scanning is the act of moving the eyes in short, abrupt, irregular changes of focus around the object of interest. The eye must stop momentarily at each point, of course, since it cannot see while moving.

(4) Factors Affecting Night Vision

(a) Lack of vitamin A impairs night vision. However, overdoses of vitamin A will not improve night vision.

(b) colds, headache, fatigue, narcotics, heavy smoking, and alcohol excess all reduce night vision.

(c) Exposure to a bright light impairs night vision and necessitates a readaptation to darkness.

(d) Darkness blots out detail. The sniper must learn to recognize objects and persons from outline alone.

b. Twilight. During dawn and dusk, the constantly changing natural light level causes an equally constant process of eye adjustment. During these periods, the sniper must be especially alert to the treachery of half light and shadow. Twilight induces a false sense of security, and the sniper must be doubly careful for his own safety. For the same reason, the enemy is prone to carelessness and will frequently expose himself to the watchful sniper. The crosshairs of the telescopic sight are visible from about one-half hour prior to sunrise until about one-half hour after sunset.

c. Illumination Aids. On occasion, the sniper may have the assistance of artificial illumination for observation and firing.

EXAMPLES:

(1) Cartridge, Illuminating, M301A2. Fixed from an 81mm mortar, this shell produces 50,000 candlepower of light which is sufficient for use of the binoculars, the spotting scope, or the rifle telescopic sight.

(2) Searchlights. In an area illuminated by searchlight, the sniper can use any of the above equipment with excellent advantage.

(3) Other. Enemy campfires or lighted areas and buildings are other aids to the observing sniper.

d. Observation Aids.

(1) Binoculars. (TA #2) Of the night observation aids, binoculars are the simplest and fastest to use. They are easily manipulated and the scope of coverage is limited only by the sniper's scanning ability. Each sniper team will be equipped with binoculars to aid in observing the enemy and in searching for and selecting targets. The binocular, M17A1, 7 x 50, has seven power magnification and a 50mm objective lens. Focal adjustments are on the eyepiece with separate adjustments for each eye. The left monocle has a horizontal and vertical scale pattern graduated in mils that is visible when the binoculars are in use.

(a) Method of Holding. Binoculars should be held lightly, monocles resting on and supported by the heels of the hands. The thumbs block out light that would enter between the eye and the eyepiece. The eyepieces are held lightly to the eye to avoid transmission of body movement. Whenever possible, a stationary rest should support the elbows.

(b) Adjustments

(1) Interpupillary Adjustment. The interpupillary distance (distance between the eyes) varies with individuals. The two monocles that make up a pair of field glasses are hinged together so that the receptive lenses can be centered over the pupils of the eyes. Most binoculars have a scale on the hinge, allowing the sniper to preset the glasses for interpupillary distance. To determine this setting, the hinge is adjusted until the field of vision ceases to be two overlapping circles and appears as a single sharply defined circle.

(2) Focal Adjustment. Each individual and each eye of that individual requires different focus settings. Adjust the focus for each eye in the following manner:

(a) With both eyes open, look through the glasses at a distant object.

(b) Place one hand over the objective lens of the right monocle and turn the focusing ring to the left monocle until the object is sharply defined.

(c) Uncover the right monocle and cover the left one.

(d) Rotate the focusing ring of the right monocle until the object is sharply defined.

(e) Uncover the left monocle; the object should then be clear to both eyes.

(f) Read the diopter scale on each focusing ring and record for future reference.

(c) Reticle. The mil scale that is etched into the left lens of the binoculars is the reticle pattern and is used in adjusting artillery fire and measuring vertical distance in mils. The horizontal scale is divided into 10-mil increments. The zero line is the short vertical line that projects below the horizontal scale between two numbers "1". To measure the angle between two objects (such as a target and an artillary burst), center the target above the zero line. Then read the number which appears on the scale under the artillery burst. There are two sets of mil scales, one above the zero on the horizontal scale, the other above the left horizontal 50-mil line on the horizontal scale. The vertical scales are divided into increments of 5 mils each. The vertical angle between the house and point A at the base of the tree is 10 mils. The third vertical scale is the range scale. It is used to estimate ranges from a known range but is not used by the sniper since he estimates his ranges by eye.

(2) Rifle Telescopic Sight, 10X Ultra. When equipped with the telescopic sight, the sniper can observe up to 800 meters with varying effectiveness in artificial illumination. In full moonlight, it is effective up to 600 meters. For best results, a supported position should be used.

(a) 10 Power. At 10 power, the field of view is more reduced and scanning clarity is impaired. High power can be used to distinguish specific objects, but scanning will lend a flat, unfocused appearance to terrain.

(3) Starlight Scope. Although the function of the starlight scope is to provide an efficient viewing capability during the conduct of night combat operations, the starlight scope does not give the width, depth, or clarity of daylight vision. However, the individual can see well enough at night to aim and fire his weapon, to observe effect of firing, the terrain, the enemy, and his own forces; and to perform numerous other tasks that confront Seal's in night combat. The starlight scope may be used by snipers to:

(a) Assist sniper teams in deployment under cover of darkness to preselected positions.

(b) Assist sniper teams to move undetected to alternate positions.

(c) Locate and suppress hostile fire.

(d) Limit or deny the enemy movement at night.

a. Factors Affecting Employment. Consideration of the factors affecting employment and proper use of the starlight scope will permit more effective execution of night operations. The degree to which these factors aid or limit the operational capabilities of the starlight scope will vary depending on the light level, weather conditions, operator eye fatigue, and terrain over which the starlight scope is being employed.

(1) Light. Since the starlight scope is designed to function using the ambient light of the night sky, the most effective operation can be expected under conditions of bright moonlight and starlight. As the ambient light level decreases, the viewing capabilities of the starlight scope diminsion. When the sky is overcast and the ambient light level is low, the viewing capabilities of the starlight scope can be greatly increased by the use of flares, illuminating shells or searchlight.

(2) Weather Conditions. Clear nights provide the most favorable operating conditions in that sleet, snow, smoke, or fog affect the viewing capabilities of the starlight scope. Even so, the starlight scope can be expected to provide some degree or viewing capability in adverse weather conditions.

(3) Terrain. Different terrain will have an adverse effect on the starlight scope due to the varying ambient light conditions which exist. It will be the sniper's responsibility to avaluate these conditions and know how each will affect his ability to observe and shoot.

(4) Eye Fatigue. Most operators will initially experience eye fatigue after five or ten minutes of continuous observation through the starlight scope. To aid in maintaining a continued viewing capability and lessen eye fatigue, the operator may alternate eyes during the viewing period.

4. Observation Telescope. The observation telescope is a prismatic optical instrument of 20-power magnification. It is carried by the sniper teams whenever justified by the nature of a mission. The lens of the telescope are coated with a hard film of magnesium flouride for maximum light transmission. This coating together with the high magnification of the telescope makes observation and target detection possible when conditions or situations would otherwise prevent positive target identification. Camouflaged targets and those in deep shadows can be distinguished, troop movements can be observed at great distances, and selective targets can be identified more readily.

a. Operation. The eyepiece cover cap and objective lens cover must be unscrewed and removed from the telescope before it can be used. The cap and cover protect the optics when the telescope is not in use. The eyepiece focusing sleeve is turned clockwise or counterclockwise until the image can be carefully seen by the operator. CAUTION: Care must be taken to prevent cross-threading of the fine threads.

(2) Observation Procedures. The sniper, having settled into the best obtainable position, is ready to search his chosen area. The process of observation is planned and systematic. His first consideration is towards the discovery of any immediate danger to himself, so he begins with a "hasty search" of the entire area. This is followed by a slow, deliberate

observation which he calls a "detailed search". Then, as long as he remains in position, the sniper maintains a constant observation of the area using the hasty and detailed search methods as the occasion requires.

(a) Hasty Search. This is a very rapid check for enemy activity conducted by both the sniper and the observer. The observer makes the search with the 7 x 50 binoculars, making quick glances at specific points throughout the area, not by a sweep of the terrain in one continuous panoramic view. The 7 x 50 binoculars are used in this type search because they afford the observer with the wide field of view necessary to cover a large area in a short time. The hasty search is effective because the eyes are sensitive to any slight movements occurring within a wide arc of the object upon which they are focused. The sniper, when conducting his hasty search, uses this faculty called "side vision" or "seeing out of the corner of the eye". The eyes must be focused on a specific point in order to have this sensitivity.

(b) Detailed Search.

(1) If the sniper and his partner fail to locate the enemy during the hasty search, they must then begin a systematic examination known as the 50-meter overlapping strip method of search. Again the observer conducts this search with the 7 x 50 binoculars, affording him the widest view available. Normally, the area nearest the sniper offers the greatest potential danger to him. Therefore, the search should begin with the terrain nearest the observer's location. Beginning at either flank, the observer should systematically search the terrain to his front in a 180-degree arc, 50 meters in depth. After reaching the opposite flank, the observer should search the next area nearest his position. This search should cover the terrain includes about ten meters of the area examined during the first search. This technique ensures complete coverage of the area. Only when a target appears does the observer use the observation scope to get a more detailed and precise description of the target. The observation scope should not be used to conduct either the hasty or detailed search as it limits the observer with such a small field of view. The observer continues searching from one flank to the other in 50-meter overlapping strips as far out as he can see.

(2) To again take advantage of his side vision, the observer should focus his eyes on specific points as he searches from one flank to the other. He should make mental notes of prominent terrain features and areas that may offer cover and/or concealment to the enemy. In this way, he becomes familiar with the terrain as he searches it.

(c) Maintaining Observation

(1) Method. After completing his detailed search, the sniper may be required to maintain observation of the area. To do this, he should use a method similar to his nasty search of the area. That is, he uses quick glances at various points throughout the entire area, focusing his eyes on specific features.

(2) Sequence. In maintaining observation of the area, he should devise a set sequence of searching to ensure coverage of all terrain.

Since it is entirely possible that this hasty search may fail to detect the initial movement of an enemy, the observer should periodically repeat a detailed search. A detailed search should also be conducted any time the attention of the observer is distracted.

OPPORTUNITY FOR QUESTIONS

SUMMARY

1. Reemphasize. During this period of instruction, we have discussed the capabilities and limitations of observation during the hours of both daylight and dark. The different techniques and aids of improving your vision were discussed.

We covered the night observation aids that are availalbe to the sniper. It was noted that the binoculars are the simplest and fastest to use. The starlight scope was discussed in detail as to it's employment and those factors affecting it's employment.

In conclusion, details of observation procedures were covered. The "Hasty Search" is understood to be the first search conducted by the sniper once he moves into position as this search is conducted to discover any immediate danger to him.

2. Remotivate. Your ability to become proficient in the techniques mentioned will allow you to see the enemy before he sees you and get that first round off.

OBSERVATION LOG

Sheet ____ of _____

Originator ______________________ Date ________________ Tour of Duty _______________

Position __________________________________ Visibility __________________________

Serial	Time	Grid Coordinate GR / Brg and Range	Event	Action or Remarks

RANGE CARD, LOG BOOK AND
FIELD SKETCHING

DETAILED OUTLINE

INTRODUCTION

1. Gain Attention. The primary mission of Seal sniper is to deliver precision fire on selected targets from concealed positions. His secondary mission is to collect information about the enemy. To do this, he must primarily be observant, first to locate prospective targets and second to be able to identify what he sees. However observant he may be, the sniper cannot be expected to exercise the sheer feat of memory necessary to remember the ranges to all possible targets within his area of observation, or to recall all tidbits of information he may come across. The means designed to assist him in this task are the range card, the log book and the field sketch.

2. Today you are going to learn how to record the data which you will need to accomplish both your primary and your secondary mission, thereby greatly enhancing your chance of achieving a first round hit and of collecting useful and useable raw intelligence.

3. Purpose.

a. Purpose. To introduce you to the range card, log book and field sketch as used by the sniper in recording range estimates and in collecting information about the enemy.

b. Main Ideas. To describe the preparation of range cards and their relationship to field sketches, and to teach how to draw adequate field sketches. Also to teach the components of various tactical reports for inclusion in the sniper log book.

4. Training Objectives. Upon completion of this period of instruction, the student will be able to:

a. Prepare a Range Card

b. Prepare a Log Book

c. Prepare a Field Sketch

BODY

1. The range card is a handy reference which the sniper uses to make rapid, accurate estimates of range to targets which he may locate in the course of his observations.

a. (Show slide #1, Field Expedient Range Card) This slide illustrates a range card which a sniper might have prepared after his arrival at a point of observation. The card is drawn freehand and contains the following information:

(1) Relative locations of dominating objects and terrain features.

(2) Carefully estimated, or map measured, ranges to the objects or features.

(3) The sniper's sight setting and holds for each range.

b. (Show slide # 2, Prepared Range Card) Prior to departure on a mission, the sniper can prepare a better range card, shown here. Upon arrival in position, he draws in terrain features and dominant objects. To avoid preparing several cards for use in successive positions, the sniper can cover a single card with acetate and use a grease pencil to draw in the area features. Copies of the prepared range card should be prepared and used whenever possible.

c. Use of the Range Card

(1) Holding. (Show slide # 2, Prepared Range Card) The sniper locates a target in the doorway of the house at 10 o'clock from his position. From his card, he quickly determines a range of 450 yards and holds at crotch level. He centers the crosshairs on the crotch, fires, and hits the target in the center of the chest.

(2) Sight Setting. The sniper locates a target on the roof of the house at one o'clock from his position. He notes the sight setting 61, applies that sight setting, and fires.

2. The Field Sketch.

a. The field sketch is a drawn reproduction of a view obtained from any given point, and it is vital to the value of a sniper's log and range card, that he be able to produce such a sketch. As is the case for all drawings, artistic ability is an asset, but satisfactory sketches can be produced by anyone, regardless of artistic skill. Practice is, however, essential and the following principles must be observed:

(1) Work from the whole to the part. Study the ground first carefully both by eye and with binoculars before attempting a drawing. Decide how much of the country is to be included in the sketch. Select the major features which will form the framework of the panorama.

(2) Do not attempt to put too much detail into the drawing. Minor features should be omitted, unless they are of tactical importance,

or are required to lead the eye to some adjacent feature of tactical importance. Only practice will show how much detail should be included and how much left out.

(3) Draw everything in perspective as far as possible.

b. Perspective. The general principles of perspective are:

(1) The farther away an object is in nature, the smaller it should appear in the drawing.

(2) Parallel lines receding from the observer appear to converge; if prolonged, they will meet in a point called the "vanishing point." The vanishing point may always be assumed to be on the same plane as the parallel lines. Thus, railway lines on a perfectly horizontal, or flat, surface, receding from the observer will appear to meet at a point on the horizon. If the plane on which the railway lines lie is tilted either up or down, the vanishing point appears to be similarly raised or lowered. (Slide # 3) Thus, the edges of a road running uphill and away from the observer will appear to converge on a vanishing point above the horizon, and if running downhill, the vanishing point will appear to be below the horizon. (Slide #4)

c. Conventional Shapes. Roads and all natural objects, such as trees and hedges, should be shown by conventional outline, except where peculiarities of shape make them useful landmarks and suitable as reference points. This means that the tendency to draw actual shapes seen should be suppressed, and conventional shapes used, as they are easy to draw and convey the required impression. Buildings should normally be shown by conventional outline only, but acutal shapes may be shown, when this is necessary to ensure recognition, or to emphasize a feature of a building which is of tactical importance. The filling in of outlines with shadowing, or hatching, should generally be avoided, but a light hatch may sometimes be used to distinguish wooded areas from fields. Lines must be firm and continuous.

d. Equipment. The sniper should have with him the following items:

(1) Suitable paper in a book with a stiff cover to give a reasonable drawing surface.

(2) A pencil, preferably a No. 2 pencil with eraser

(3) A knife or razor blade to sharpen the pencil

(4) A protractor or ruler, and

(5) A piece of string 15" long.

e. Extent of Country to be Included. A convenient method of making a decision as to the extent of the country to be drawn in a sketch is to hold a protractor about 11 inches from the eyes, close one eye, and consider the section of the country thus covered by the protractor to be the area sketched. The extent of this area may be increased or decreased by moving the protractor nearer to, or farther away from, the eyes. Once the most satisfactory distance has been chosen, it must be kept constant by a piece of string attached to the protractor and held between the teeth.

f. Framework and Scale. The next step is to mark on the paper all outstanding points in the landscape in their correct relative positions. This is done by noting the horizontal distance of these points from the edge of the area to be drawn, and their vertical distance above the bottom line of this area, or below the horizon. If the horizontal length of the sketch is the same size as the horizontal length of the straight edge of the protractor, the horizontal distances in the picture may be gotten by lowering or raising the protractor and noting which graduations on its straight edge coincide with the feature to be plotted; the protractor can then be laid on the paper and the position of the feature marked against the graduation noted. The same can be done with vertical distances by turning the straight edge of the protractor to the vertical position.

g. Scale. The eye appears to exaggerate the vertical scale of what it sees, relative to the horizontal scale, i. e. things look taller than they are. It is preferable, therefore, in field sketching to use a larger scale for vertical distances than for horizontal, in order to preserve the aspect of things as they appear to the viewer. A suitable exaggeration of the vertical scale relative to the horizontal is 2:1, which means that every vertical measurement taken to fix the outstanding points are plotted as read.

h. Filling in the Detail. When all the important features have been plotted on the paper in their correct relative positions, the intermediate detail is added, either by eye, or by further measurement from these plotted points. In this way, the sketch will be built up upon a framework. All the original lines should be drawn in lightly. When the work is completed, it must be examined carefully and compared with the landscape, to make sure that no detail of military significance has been omitted. The work may now be drawn in more firmly with darker lines, bearing in mind that the pencil lines should become darker and firmer as they approach the foreground.

i. Conventional Representation of Features. The following methods of representing natural objects in a conventional manner should be borne in mind when making the sketch:

(1) Prominent Features. The actual shape of all prominent features which might readily be selected as reference points when describing targets, such as oddly shaped trees, outstanding building, towers, etc...should be shown if possible. They must be accentuated with an arrow and a line with a description, e. g. Prominent tree with large withered branch.

(2) Rivers. Two lines diminishing in width as they recede should be used.

(3) Trees. These should be represented by outline only. Some attempt should be made to show characteristic shape of individual trees in the foreground.

(4) Woods. Woods in the distance should be shown by outline only. In the foreground, the tops of individual trees may be indicated. Woods may be shaded, the depth of shadowing becomes less with distance.

(5) Roads. Roads should be shown by a double continuous line diminishing in width as it recedes.

(6) Railways. In the foreground, railways should be shown by a double line with small cross lines (which represent the ties) to distinguish them from roads; in the distance, they will be indicated by a single line with vertical ticks to represent the telegraph poles.

(7) Churches. Churches should be shown on outline only, but care should be taken to denote whether they have a tower or a spire.

(8) Towns and Villages. Definite rectangular shapes denote houses; towers, factory chimneys and prominent buildings should be indicated where they occur.

(9) Cuts and Fills. These may be shown by the usual topographic symbols, ticks diminishing in thickness from top to bottom, and with a firm line running along the top of the slope in the case of a cut.

(10) Swamps and Marshland. They may be shown by the conventional topographic symbols.

j. Other Methods of Field Sketching.

(1) The Grid Window. A simple device which can help a great deal in field sketching can be made by taking a piece of cardboard or hard plastic and cutting out of the center of it, a rectangle 6" x 2". A piece of clear plastic sheeting or celluloid is then pasted over the rectangle. A grid of squares of ½" size is drawn on the plastic sheeting. You now have a ruled plastic window through which the landscape can be viewed. The paper on which the drawing is to be made is ruled with a similiar grid of squares. If the frame is held at a fixed distance from the eye by a piece of string held in the teeth, the detail seen can be transferred to the paper square by square.

(2) The Compass Method. Another method is to divide the paper into strips by drawing vertical lines denoting a fixed number of mils of arc and plotting the position of important features by taking compass bearings to them. This method is accurate but slow.

3. The Sniper Log Book. The sniper log is a factual, chronological record of his employment, which will be a permanent source of operational data. It will provide information to intelligence personnel, unit commanders, other snipers and the sniper himself. Highly developed powers of observation are essential to the sniper, as explained in earlier lessons. Because of this, he is an important source of intelligence whose reports may influence future operations, and upon which many lives may depend.

a. Data to be Recorded. The log will contain at a minimum, the following information:

(1) Names of observers,
(2) Hours of observation,
(3) Data and Position (Grid Coordinates)
(4) Visibility
(5) Numbered observations in chronological order,
(6) Time of observation,
(7) Grid reference of observation,
(8) Object seen, and
(9) Remarks or action taken.

b. Supplementary Materials. The sniper log is always used in conjunction with the field sketch. In this way, not only does the sniper have a written account of what he saw, but also a pictorial reference showing exactly where he sighted or suspected enemy activity. If he is then relieved in place, the new observer can more easily locate earlier sightings, by comparing the field sketch to the landscape, than he could solely by use of grid coordinates.

OPPORTUNITY FOR QUESTIONS AND COMMENTS

SUMMARY

1. Reemphasize. During this period of instruction, we have looked at the Range Card, The Field Sketch, and The Log Book. Each was presented as an entity in itself and as how each related to the other.

2. Remotivate. The primary and secondary missions of the sniper were mentioned at the beginning of this class. The primary mission can be fulfilled without use of a log book and the secondary mission can be accomplished without recourse to a range card. However, to be a complete Seal Scout/Sniper, you must be able to prepare a useable field sketch and range card to ensure accurate shooting and a thorough log book to collect all available intelligence data.

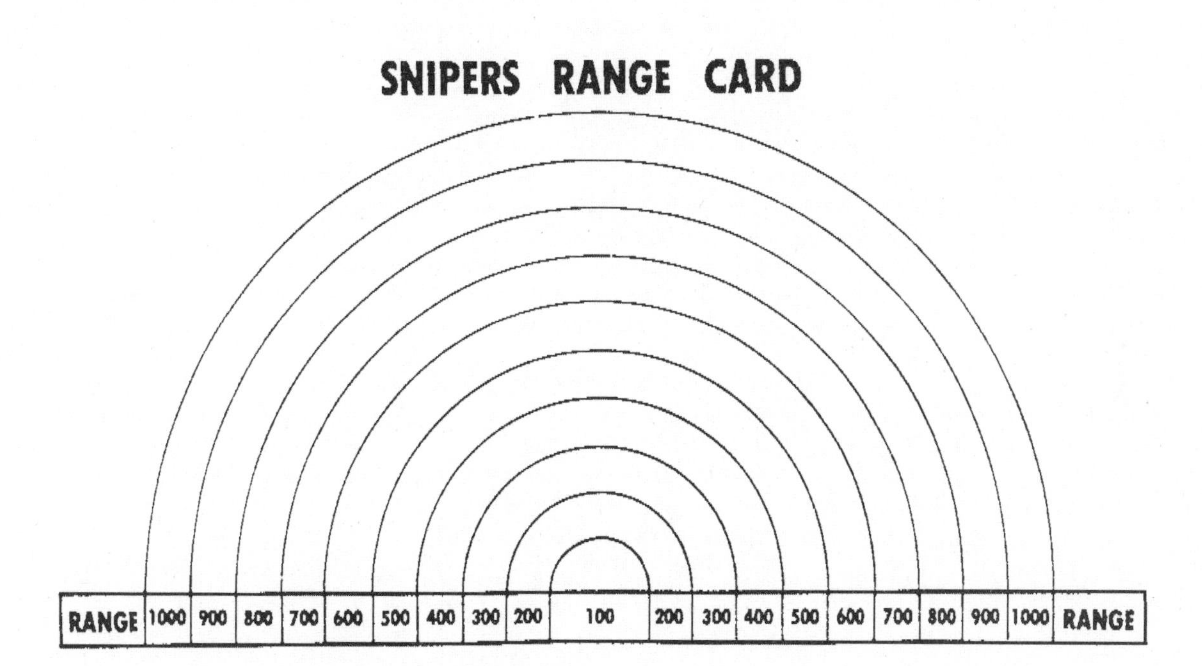

GRID COORDINATE OF POSITIONS ______________________

METHOD OF OBTAINING RANGE ______________________

MADE OUT BY ______________________

DATE ______________________

SIGHTING, AIMING AND TRIGGER CONTROL

INTRODUCTION

1. Gain Attention. The first of the basic marksmanship fundamentals taught to the shooter are sighting, aiming and trigger control. The reason for this is that without an understanding of the fundamentals, a sniper will not be able to accomplish his primary mission.

2. Motivate. During your eight weeks here at the school, you will be taught individual movement, cover and concealment, map reading and many other related sniper skills, in addition to marksmanship. By the time you graduate, you will be able to move to a sniper position, fire a shot and withdraw, all without being observed. However, all this will be to no avail if, because you do not understand the principles of sighting, aiming and trigger control, you miss your target when you do fire.

3. State Purpose and Main Ideas.

A. Purpose. To introduce the student to the principles of sighting, aiming and trigger control as applied to the sniper rifle with telescopic sight.

B. Main Ideas.

(1) Sighting and aiming will be discussed in three phases:

(a) The relationship between the eye and the sights

(b) Sight picture

(c) Breathing

(2) Trigger control will be explained through smooth action, interrupted pull, concentration and follow-through.

4. Training Objectives. Upon completion of this period of instruction, the student will be able to:

A. Understand and demonstrate the sighting error known as "shadow effect";

B. Understand and demonstrate "Quartering" the target.

C. Understand and demonstrate proper trigger control.

TRANSITION. The arrangement of an optical sight allows for aiming without recourse to organic rifle sights. The role of the front sight in a telescope is fulfilled by the crosshairs. Because of the crosshairs and the image of the target are in the focal plane of the lens, the shooter can use both at the same time and with equal clarity.

BODY

A. Aiming

(1) Relationship between the eye and the sights. In order to see what is required during aiming, the shooter must know how to use his eye. Variations in the positions of the eye to the telescope will cause variations in the image received by the eye. The placement of eye in this respect is called eye relief. Proper eye relief is approximately 2-3 inches from the exit pupil of the telescope, and can be determined to be correct when the shooter has a full field of view in the telescope with no shadows. If the sniper's eye is located without proper eye relief, a circular shadow will occur in the field of vision, reducing the field size, hindering observation, and, in general, making aiming difficult. If the eye is shifted to one side or another of the exit pupil, cresent shaped shadows will appear on the edges of the eyepiece (TP #1 Shadow Effects). If these cresent shaped shadows appear, the bullet will strike to the side away from the shadow. Therefore, when the sniper has a full field of view and is focusing on the intersection of the crosshairs, he has aligned his sight.

(2) Sight Picture. With the telescopic sight, this is achieved when the crosshairs are centered on the target, and the target has been quartered. (Place on TP #3, Sample Sight Pictures). This transparency shows samples of different types of sight pictures. In each case, you can see that the target has been quartered to maximize the chance of a first round hit.

(3) Breathing. The control of breathing is critical to the aiming process. If the sniper breathes while aiming, the rising and falling of his chest will cause the muzzle to move vertically. To breath properly during aiming, the sniper inhales, then exhales normally and stops at the moment of natural respiratory pause. The pause can be extended to 8-10 seconds, but it should never be extended until it feels uncomfortable. As the body begins to need air, the muscles will start a slight involuntary movement, and the eyes will loose their ability to focus critically. If the sniper has been holding his breath for more than 8-10 seconds, he should resume normal breathing and then start the aiming process over again.

B. Trigger Control

(1) The art of firing your rifle without disturbing the perfected aim is the most important fundamental of marksmanship. Not hitting where you aim is usually caused by the aim being disturbed just before or as the bullet leaves the barrel. This can be caused by jerking the trigger or flinching as the rifle fires. A shooter can correct these errors by using the correct technique of trigger control.

(2) Controlling the trigger is a mental process, while pulling the trigger is a physical one. Two methods of trigger control used with the

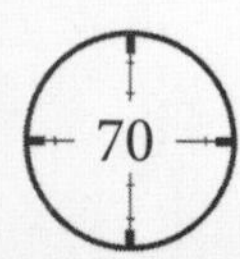

(a) Using the smooth action method, the shooter takes up the initial pressure, or free play, in the trigger. Then, when the aim is perfected, increases the pressure smoothly until the rifle fires.

(b) When using the interrupted method of trigger pull, the shooter takes up the initial pressure and begins to squeeze off the shot when the aim is perfected. However, because of target movement or weapon movement, he pauses in his trigger squeeze until the movement stops, then continues to squeeze until the weapon fires.

(c) Trigger Control Developed as a Reflex. The shooter can develop his trigger control to the point that pulling the trigger no longer requires conscious effort. The shooter will be aware of the pull, but he will not consciously be directing it. A close analogy to trigger control can be found in typing. When first learning to type, the typist reads the alphabetic letter to be typed, mentally selects the corresponding key, and consciously directs a finger to strike the key. After training and practice, however, the typist will see the letter which has to be typed and the finger will hit it automatically. This then, is a conditioned reflex; conditioned because it was built in and reflex because it was not consciously directed.

The same type of conditioned reflex can be developed by the sniper. When he first starts firing, he must consciously direct his finger to squeeze the trigger as soon as the aim is perfected. As a result of training, however, a circuit will be established between the eye and the trigger finger. The eye, seeing the desired sight picture, will cause the finger to squeeze the trigger without conscious mental effort. The shooter, like the typist is aware of pressure against the trigger, but is not planning or consciously directing it.

(d) Developing Trigger Control. In all positions, one of the best methods for developing proper trigger control is through dry firing, for here the shooter is able to detect his own errors without having recoil conceal undesirable movements. Only through patience, hard work, concentration and great self-discipline will the mastery of trigger control be achieved.

(3) Factors Affecting Trigger Control.

(a) Concentration. The shooter's concentration should be focused on the perfection of aim, as trigger control is applied. Concentration defined as the will to demand obedience, is the most important factor in the technique of trigger control.

(b) Placement of the Trigger Finger and Grip on the Rifle. The finger should be placed on the trigger in the same place each time. Only through practice can the shooter determine which part of his finger should go on which spot on the trigger. Any position of the finger in relation to the trigger is acceptable so long as the shooter can pull the trigger straight to the rear.

Moreover, in order to achieve a smooth, consistent trigger squeeze, the stock must be grasped firmly and in the same place each time.

(c) Follow-Through. Follow-through means doing the same things after each shot is fired, thereby insuring that there is no undue movement of the rifle before the bullet leaves the barrel. The shooter continues to hold his breath, to focus on the crosshairs and to practice trigger control even though the rifle has already fired. By doing this, the shooter can detect any errors in sight alignment and sight picture and he can correct them after follow-through has been completed.

OPPORTUNITY FOR QUESTIONS AND COMMENTS (1 MIN)

SUMMARY (1 MIN)

1. Reemphasize. We have just covered the marksmanship fundamentals of sighting, aiming and trigger control, and how to apply them properly.

2. Remotivate. Your ability to hit a target at any range and under any conditions will be a measure of how well you have practiced and mastered these principles.

CORRECTING FOR ENVIRONMENTAL FACTORS

1. PURPOSE. In the case of the highly trained sniper, effects of the weather are a primary cause of error in the strike of the bullet. The wind, mirage, light, temperature, and humidity all have effects on the bullet, the sniper, or both. Some effects are insignificant depending on the average conditions of sniper employment. However, sniping is accomplished under extreme of weather and therefore all effects must be considered. The observation telescope will not only assist you in detecting well camouflaged targets, it may also be used to read mirage. This two hour conference will provide you with the necessary information to compensate for the effects of wind and weather and to properly use the observation telescope.

2. OBJECTIVES:

a. Training Objective. To enable the sniper to explain the effects of wind and weather on the strike of the bullet and how to compensate for these effects and use the observation telescope in accordance with TC 23-14.

b. Training Objective. As a student in a field environment, accomplish the following training objectives in accordance with TC 23-14 and as discussed in class:

(1) Explain the effects of wind.

(2) Explain the effects of mirage.

(3) Explain the effects of temperature.

(4) Explain the effects of light.

(5) Explain the effects of humidity and altitude.

(6) Use the observer's telescope.

The objective of this lesson was to enable you to explain the effects of wind and weather on the strike of the bullet and how to compensate for these effects, and to use the observation telescope in accordance with TC 23-14.

a. The conditions which constantly present the greatest problem to the sniper is the wind. The flag, drop field expedient, or clock themth may be used to estimate wind velocity. However, the most accurate method of determining wind velocity is by eading mirage.

b. To determine wind velocity using the flag (or drop) method, divide the angle the range flag (or light object dropped from the shoulder) makes from the verticle by a constant of 4.

c. To determine the number of clicks necessary to compensate from the wind, using either iron sight or scope, the following formulas are used:

(1) For M118 NM ammunition use:

$\frac{RxV}{10}$ = Number of clicks for a full value wind

(2) For M80 ball ammunition use:

$\frac{RxV}{15}$ = Minutes of angle for a full value wind

NOTE: For rantes less than or equal to 500 yards; beyond 500 yards constant 15 changes to bullet velocity loss.

RANGE	CONSTANT
600	14
700	13
800	13
900	12
1,000	11

d. Using M80 ball ammunition, the answer in minutes of angle must be multiplied times 2 to determine clicks.

e. For a half value wind, divide <u>clicks by 2</u>.

f. To determine the amount of "hold-off" needed to compensate for the wind the following formula is used.

$\frac{C \times R}{2}$ = Inches of hold-off

C: Is the number of clicks.
R: Is the range to the target in nearest hundreds(i.e., 500 yards, converts to 5).
V: Is the wind velocity in MPH.

g. Field expedient method:

WIND	EFFECTS
01-03	Direction of wind shown by smoke but not by flag.
04-07	Wind felt on face, leaves rustle.
08-12	Leaves and small twigs in constant motion.
13-18	Raises dust and loose paper; small branches are moved.
19-24	Small tree in leaf being to sway.

h. There are three types of mirage:

SLOW	FAST	BOILING
0-7 MPH	8-15 MPH	No value

Mirage also has density, in addition to direction:

THICK	THIN
High Temperature, High Humidity, Bright Light	Low Temperature and/or Low Light

i. To read mirage, the sniper/observer focuses his spotting scope at mid-range.

(1) At 300 yards, a +20 degree change in temperature necessitates a +1 minute of angel change. *

(2) At 600 yards, a +15 degree change in temperature necessitates a +1 minute of angle change. *

(3) At 1,000 meters, a 10 degree change in temperature necessitates a +1 minute of angle change. *

*NOTE: Remember that 1 minute of angle = 2 clicks.

j. The rule of thumb for temperature correction is temperature up, sights down and temperature down, sights up.

k. The rule of thumb for humidity is humidity up, sights up and humidity down, sights down.

l. The rule of thumb for changes in light conditions is lights up, sights up and light down, sights down.

m. Elevation above sea level can have an important effect on bullet trajectories. At higher elevations, both air density and temperature decreases, and air drag on the bullet decreases. At higher elevations, the sniper has a tendency to shoot high.

n. The observer's telescope not only aids the sniper in detecting targets, it is also a valuable tool in reading mirage. The observer telescope, when reading mirage, is focused at mid-range.

APPLICATION OF FIRE

1. PURPOSE: Under normal conditions, all sniping occurs over unknown distances. Without a thorough grounding in the practical application of external ballistics, it is unlikely that a sniper will be capable of hitting his targets at any point but the shortest ranges. This lesson concerns itself with the fundamentals of unknown distance shooting and the application of exterior ballistics.

2. OBJECTIVES:

a. Objective. To enable the sniper to apply the fundamentals of exterior ballistics in the engagement of targets at unknown distances to include; definition and application of minute of angle corrections, minute of angle conversions, indication of targets and fire control orders, and practical application of exterior ballistics for firing over ground and computing hit probability.

b. Training Objectives. As a sniper demonstrate the application of the following training objectives in accordance with FMFM 1-3B and TC 23-14.

(1) Calculate minute of angle corrections and conversions.

(2) Define the components and the factors which influence a bullet's trajectory.

(3) Demonstrate the indication of targets at unknown distances.

APPLICATION OF FIRE

1. A minute of angle (MOA) is an angular measure which subtends 1/60th of one degree of arc; and for practical purposes is the equivalent of one (1) inch per 100 yards of range, ie, 1 MOA = 3 inches at 300 yards, or 3 cm per 100 meters of range, ie, 5 MOA = 75 cm at 500 meters.

2. To determine minutes of correction, divide the error in inches or centimeters by the whole number of the range in hundreds of yards or meters.

$$\text{MINUTTES} = \frac{\text{ERROR}}{\text{RANGE}}$$

3. To convert mils to minutes of angle, multiply minutes mils by 3.375.

1 Mil = 3.375 Minutes

4. Trajectory is the path a bullet follows when fired from a weapon.

5. The factors which influence trajectory are:

   a. The initial (muzzle) velocity.

   b. The angle of departure.

   c. Air resistance.

   d. The rotation of the projectile about its axis.

   e. Gravity.

6. Angle of departure is the elevation in minutes or degrees that must be imparted to the barrel through sight corrections, in order that the bullet will strike a target at a specific distance.

7. Angle of departure, and therefore point of impact, is not constant and is affected by four (4) variables:

   a. Variations in initial velocity due to imperfections of ammunition.

   b. Imperfections in aiming.

   c. Imperfections in the rifle.

   d. Errors in holding and canting the rifle.

8. Air resistance is the most significant factor in trajectory.

9. Maximum ordinate is the highest point of the trajectory. It divides the trajectory into the rising branch and the falling branch.

10. Point of aim is the point where the line of sight meets the target.

11. Point of impact is the point where the bullet strikes the target.

12. There are two (2) danger spaces:

a. The distance in front of the muzzle, within which the bullet does not rise higher than the object fired at, is called the danger space of the rising branch.

b. The distance beyond the maximum ordinate, within which the bullet drops from the height of the target to below target level, is the danger space of the falling branch. This is divided into the danger space in front of the target, and danger space behind the target, with the height of the point of aim as the dividing point.

13. The extent of the danger space is dependent upon:

a. Height of the firer.

b. Height of the target.

c. The flawlessness of the trajectory.

d. The angle of the line of sight.

e. The slope of the ground where the target resides.

14. There are three (3) methods of indicating targets:

a. Direct method.

b. Reference point method.

c. Clock ray method.

15. When indicating targets, the following information must be given:

a. Range.

b. Corrections - For wind, leads, or hold off.

c. Indications.

16. The sniper team must learn to work together in such a manner so that each knows exactly what the other means in as few words as possible. Before the sniper fires he must ensure that the observer is ready to "read" the shot.

17. The sniper team, before they start a mission, must agree and understand what methods will be used to indicate targets to the other and what methods will be used to ensure both are ready before the shot is made.

18. When firing over ground, the extent of its danger space depends on the relationship between:

a. The trajectory and the line of sight, or angle or fall, and therefore on the range and the circumstances of its trajectory.

b. On the height of the target.

c. On the point of aim.

d. On the point of impact.

20. For the given height of target and point of aim, the danger space is of fixed dimensions over level ground, while the swept space varies in relationship to the slope of the ground; being greater on falling ground and lesser on rising ground.

The following is a list of compensation factors to use in setting the sights of the sniper weapon system when firing from any of the following angles. To use this table, find the angle at which you must fire and then multiply the estimated range by the decimal figure shown to the right, i.e. estimated range is 500 meters, angle of fire is 35 degrees, set zero of weapon for:

500 * .82 = 410 meters

SLOPE ANGLE UP OR DOWN	MULTIPLY RANGE BY
.05 deg.	.99
.10 deg.	.98
.15 deg.	.96
.20 deg.	.94
.25 deg.	.91
.30 deg.	.87
.35 deg.	.82
.40 deg.	.77
.45 deg.	.70
.50 deg.	.64
.55 deg.	.57
.60 deg.	.50
.65 deg.	.42
.70 deg.	.34
.75 deg.	.26
.80 deg.	.17
.85 deg.	.09
.90 deg.	.00

As can quickly be seen the steeper the angle the shorter the range will be set on the scope or sights to cause a first round hit. Also the steeper the angle the more precise you must be in estimating or measuring the angle. Interpolation is necessary for angles between tens and fives.

As an example 72 degrees is 40 % between 70 and 80 degrees
70 degrees = .34 and 75 degrees = .26
(.34 + .26) / 2 = .30 or 72.5 degrees - .30
72 degrees would equal approximately .31
A range of 650 meters at a 72 degree angle would equal:
650 * .31 = 201.5 meter zero.

Interpolation can be further carried out to 71 degrees or 74 degrees by using the same method with the .30 found for 72.5 degrees:

(.30 + .26) / 2 = .28 for 73.75 degrees or (.30 + .34) / 2 = .32 for 71.25 degrees

LEADS

INTRODUCTION

1. Gain Attention. You and your partner have been in position for several days without any luck at all, and are just packing it in when your partner catches sight of someone moving down a dry river bed, approximately 675 to 700 yards down range. You both decide that he is moving at about a 45 degree angle to you, an at average pace. You obtain what you think is the proper hold and lead for that range and squeeze the shot off. Your partner doesn't say anything, but looks at you and winks.

2. First round kill is the name of the game. Being snipers, you could very well be placed in this situation and when you are, will be expected to put that round right where it belongs on a moving target out to 800 yards.

3. Purpose.

a. Purpose. The purpose of this period of instruction is to provide the student with the knowledge of the proper leads to be used to hit a moving target (walking and running) at ranges from 100 to 800 yards.

b. Main Ideas. The main ideas to be discussed are the following:

(1) Methods of Leading a Moving Target
(2) Angle of Target Movement
(3) Normal Leads
(4) Double Leads

4. Training Objectives. Upon completion of this period of instruction, the student will, without the aid of references, understand and be able to demonstrate the proper lead necessary to hit a moving target at ranges from 100 to 800 yards.

TRANSITION. The best example of a lead can be demonstrated by a quarterback throwing a pass to his receiver. He has to throw the ball at some point down field to where the receiver has not yet reached. The same principle applies in shooting at a moving target with the sniper rifle.

BODY

1. LEADS. Moving targets are the most difficult to hit. When engaging a target which is moving laterally across his line of sight, the sniper must concentrate on moving his weapon with the target while aiming at a point

some distance ahead. Holding this "lead", the sniper fires and follows through with the movement after the shot. Using this method, the sniper reduces the possibility of missing, should the enemy suddenly stop, hit the deck, or change direction. The following is a list of ranges and leads used to hit moving targets both walking and running;

	WALKING	RUNNING
RANGE	LEAD	LEAD
100	Front edge of body	½ foot/body width
200	½ foot/body width	1 foot/body width
300	1 foot/body width	2 feet/body width
400	1½ feet/body width	3 feet/body width
500	2 feet/body width	4 feet/body width
600	2½ feet/body width	5 feet/body width
700	3 feet/body width	6 feet/body width
800	3½ feet/body width	7 feet/body width

Another method of leading a target, and one which is used extensively by the British, is known as the "Ambushing". By "ambush", we mean the sniper selects a point some distance in front of his target and holds the cross-hairs or Mil Dots on that point. As the target moves across the horizontal crosshair or Mil Dot, it will eventually reach a point which is the proper lead distance from the center. At that instance, the sniper must fire his shot. This is a very simple method of hitting a moving target, but a few basic marksmanship skills must not be forgotten:

a. The sniper must continue to concentrate on the crosshairs and not on the target.

b. The sniper must continue to squeeze the trigger and not jerk or flinch prior to the shot being fired.

c. Some snipers tend to start with this method, but begin to track the target once it reaches that magic distance and then fire the shot. Use one of the two methods and stick with the one which you are confident will get that shot on target. (The instructor should draw these methods of leading on the chalkboard to better illustrate.)

TRANSITION. The sniper must not only estimate his target range, but also it's speed and angle of travel relative to his line of sight in order to determine the correct lead.

(a) Full Lead Target. When the target is moving across the observer's front and only one arm and one side are visible, the target is moving at or near an angle of 90 degrees and a full value lead is necessary.

(b) Half Lead Target. When one arm and two-thirds of the front or back are visible, the target is moving at approximately a 45 degree angle and a one-half value lead is necessary.

(c) No Lead Target. When both arms and the entire front or back are visible, the target is moving directly toward or away from the sniper and will require no lead.

OPPORTUNITY FOR QUESTIONS

SUMMARY

1. Reemphasize. During this period of instruction, we have discussed the two different methods most often used to lead a moving target and emphasized that it was important to stick with one method and not fluctuate back and forth between the two.

We covered the required leads that should be used to hit a moving target out to 800 yards.

In conclusion, we discussed how to estimate angle of target movement and use of a full lead and half lead. Double leads were covered and the situation was covered as when to apply them.

2. Remotivate. As you can see, the sniper must now become proficient in his ability to judge distance, how fast his target is moving, and at what angle the target is moving with respect to him and still put that first round on target at ranges out to 800 yards.

MOVING TARGETS

SCHEDULE

Dates	Times	Location / Yd Line	Rounds Needed
		R4 / 300 - 600 Yd Line	
		R4 / 300 - 600 Yd Line	
		R4 / 700 and 800 Yd Line	
		R4 / 700 and 800 Yd Line	
		R4 / 700 and 800 Yd Line	
		R4 / 300 and 800 Yd Line	
		R4 / 300 and 800 Yd Line	

REQUIREMENTS

1. Twelve (12) 12" FBI silhouette targets for all the above dates.

2. Seven (7) "A" type targets.

3. In addition to normal combat equipment, each sniper team will be equipped with sniper rifle and binoculars/Spotting Scope.

4. Student uniforms as directed.

1. State Purpose and Main Ideas.

a. Purpose. To make the sniper determine the proper leads necessary to hit a target walking or running at ranges of 100 to 800 yds.

b. Main Ideas. The main ideas to be discussed are the following:

(1) Methods of leading a moving target.
(2) Angle of target movement.
(3) Speed of target.
(4) Normal Leads.
(5) Double Leads.

2. Training Objectives. Upon completion of this period of instruction, the student will:

a. Be able to understand the proper methods or leading a target at ranges of 100-800 yds.

BODY

1. Conduct of Firing Exercise.

a. The student will wear camouflage and move tactically during the moving exercise. Tactical movement for this exercise will consist of a low crawl from behind the firing point to the firing point (approximately 10-15 yds) and back.

b. Each sniper team will be given a block of targets that will be his firing position (approximately 35 ft long). Each sniper team must pick a firing position within his sector of fire and low crawl to his firing position.

c. The student then must load five (5) rounds of ammunition into his sniper rifle and wait until a target appears in his sector of fire.

d. Moving targets will appear on the far left sector first and far right sector second of the snipers firing sector.

e. When a target appears, the sniper's observer must tell the sniper where in the sector the target is, the wind element at the time of the sighting and any other element that may cause a error in a first round hit.

f. After the sniper has engaged his target and it is a hit, the target will go down and move to the far corners of their sector of fire and wait until all targets have reached this position. On command from the Pit Officer or OIC, all targets will be sent into the air and show their hits. The sniper's observer will then record his hold used and plot the impact of the shot.

g. If a target has reached the end of the sector and it has not been fired upon or hit, the student will bring the target into the pits. On command from the Pit Officer/OIC all targets will go into the air. A miss will be indicated by facing the back side of the target towards the firing line. On the command from the Pit Officer/OIC all targets will be taken back into the pits, and again on command from the Pit Officer/OIC, the next set of targets will come up and start to move from right to left in their sector of fire.

2. Conduct of Pit Officer/OIC.

a. To insure targets, pasters, and spotters are available for each target.

b. To insure all targets are spaced:

(1) Pit team 1 - 1-8
(2) Pit team 2 - 9-16

(3) Pit team 3 - 17-24
(4) Pit team 4 - 25-32
(5) Pit team 5 - 33-40
(6) Pit team 6 - 41-50

c. To insure that two (2) students are manning each moving targets.

d. To contact the Conducting Officer/OIC when pits are ready to start the firing exercise.

e. On command from the Conducting Officer/OIC each student will raise a moving target at the far left sector of fire, (e. g. 1, 9, 17, 25, 33, 41) and start walking from left to right or to the end of the firing sector (e. g. 8, 16, 24, 32, 40, 50).

f. The Pit Officer/OIC must insure that all targets start at the far left sector of fire first. (e. g. targets 1, 9, 17, 25, 33, 41)

g. On command, raise all moving targets and walk to the end of the sector (e. g. 8, 16, 24, 32, 40, 50).

h. Insure that if a target is hit, the student pulls the target into the pits and walks to his far sector or end of his sector and waits. If a target is not engaged or the sniper fires and misses, insure that the target keeps moving until it reaches the end of the sector and then brought down into the pits.

i. On command from the Conducting Officer/OIC, all targets will appear to show the student their impact. If it is a miss, the back side of the moving target will appear. On command all targets will be taken back into the pits.

j. Again on command the next set of targets will start to move from right to left.

k. If the student hears the word "mark" in their sector of fire, he will pull the target down and look for a shot. If an impact hole can not be found, raise the moving target and walk to the end of the sector.

3. Scoring.

a, The value of each hit will be determined by the formula
V = R / 100; V = value of the hit, R = the range
i.e. V = 500/100 = 5
V = 600/100 = 6
V = 300/100 = 3

b. Misses will be scored as zero.

c. Passing score for a firing exercise will be 80% of the total points available.

FORMALA TO DETERMINE WIND CORRECTIONIN MILS

1. DETERMINE WIND CORRETION

RANGE X VELOCITY = WIND CORRECTION IN M.O.A.
15

14 600 yds
13 700 yds
13 800 yds
12 900 yds
11 1000 yds

2. CONVERT NORMAL MIL LEAD INTO M.O.A.

LEAD IN MILS X 3.375 = LEAD IN M.O.A.

3. SUBTRACT NORMAL LEAD IN MILS CONVERTED TO M.O.A. FROM WIND CORRETION IN M.O.A.

LEAD IN M.O.A. - WIND CORRETION IN M.O.A. = CORRECT LEAD IN M.O.A.

4. TO DETERMINE NEW LEAD IN MILS ÷ CORRECT LEAD IN M.O.A.

3.375 = LEAD IN MILS divided by NEW LEAD IN M.O.A.

EXSAMPLE.

500 yard shot with 8 m.p.h. full value wind from the right.

1. (5 x 8) = 2.6 2. 1 1/4 x 3.375 = 4.75 MOA 3. 4.75 - 2.6 = 2.15 MOA
15

4. 3.375 ÷ 2.15 = 1.5 MILS

MOVING TARGET LEADS

NOTE:HOLDS ARE FROM CENTER OF TARGET

WALKING-2 MPH

RANGE	MILS	MOA	FEET
100 yards	Leading edge of target.		
200 yards	1	3	0.5
300 yards	1 1/8	4	1.0
400 yards	1 1/4	4.5	1.5
500 yards	1 1/4	4.75	2.0
600 yards	1 1/2	5.0	2.5
700 yards	1 1/2	5.0	3.0
800 yards	1 1/2 (1 3/4)	5.25 (5.5)	3.5 (3.75)
900 yards	1 1/2 (1 3/4)	5.25 (5.5)	4.0 (4.25)
1000 yards	1 3/4	5.5 (5.75)	4.5 (5.0)
	RUNNING SHELL		
100 yards	1 3/4	6	0.5
200 yards	1/34	6	1.0
300 yards	2 1/4	8	2.0
400 yards	2 1/2	9	3.0
500 yards	* 2 3/4	9.5	4.0
600 yards	* 3	10	5.0
700 yards	* 3	10.25	6.0
800 yards	* 3	10.5	7.0
900 yards	* 3 1/4	10.75	8.0
1000 yards	* 3 1/4	10.75	9.0

*Running targets are not recommended at these ranges due to the leads required. If target must be engaged then use windage knob and hold directly on the leading edge of the target. Example, runner left to right, 600 yards, use right 10 minutes of angle and hold on his leading edge.

FAST WALKERS-3-4 MPH

RANGE	MILS	MOA	FEET
100 yards	Leading edge of the target		
200 yards	1 1/4	4.5	0.75
300 yards	1 3/4	6.0	1.75
400 yards	2	6.5	2.5
500 yards	2	6.5	3.0
600 yards	2	6.75	4.0
700 yards	2	6.75	4.5
800 yards	2 1/4	8.0	5.5
900 yards	2 1/4	8.0	6.75
1000 yards	2 1/2	8.5	7.5

It must be emphasized that these are beginning leads only. Each individual will have his own lead for any given time and/or circumstances. The wind will also play a very big factor in the lead used for a given shot at a given range. As an example, a walker at 600 yards moving from left to right with a wind of 15 mph from left to right will change the lead from 2.5 feet or 1.5 mils to the leading edge of the target. (15 * 6 = 90/10 = 9/2 = 4.5 MOA. Normal lead for a walker at 600 yards is 5 MOA, a difference of .5 MOA or 3 inches at 600 yards.)

Half leads for angular movement and double leads for movement towards the shooters shooting hand must also be computed into the lead.

.224 dia., 55 gr. FULL METAL JACKET B.T.

RANGE YARDS		MUZZLE	100	200	300	400	500
VEL. FPS		2900	2545	2187	1860	1453	1139
ENERGY FT. LB.		1027	791	584	422	258	158
DROP INCHES		00	-2.22	-9.85	-24.81	-50.13	-92.26
WIND DEFLECTION INCHES	10 MPH	00	1.15	5.33	13.31	27.05	50.21
	20 MPH	00	2.31	10.65	26.61	54.10	100.42
	30 MPH	00	3.46	15.98	39.92	81.16	150.63

.308 dia., 150 gr. FULL METAL JACKET B.T.

RANGE YARDS		MUZZLE	100	200	300	400	500
VEL. FPS		2800	2597	2404	2218	2041	1872
ENERGY FT. LB.		2611	2247	1924	1638	1387	1167
DROP INCHES		00	-2.30	-9.76	-23.24	-43.80	-72.72
WIND DEFLECTION INCHES	10 MPH	00	.72	3.00	7.01	12.97	21.13
	20 MPH	00	1.45	6.00	14.02	25.94	42.25
	30 MPH	00	2.17	8.99	21.03	38.91	63.38

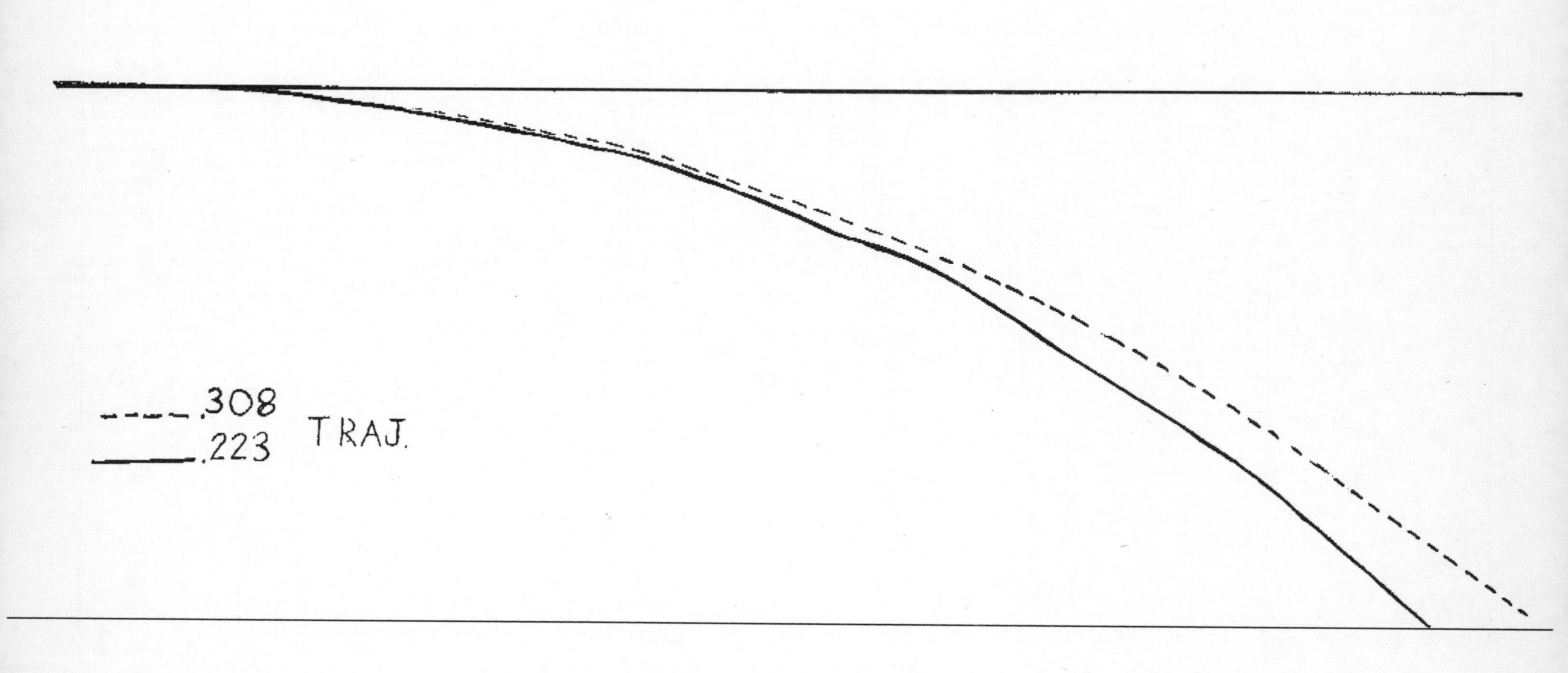
.308
.223
TRAJ.

VITAL ZONE

150 GR FMJ .308 @ 2800 FPS
55 GR FMJ .223 @ 2900 FPS

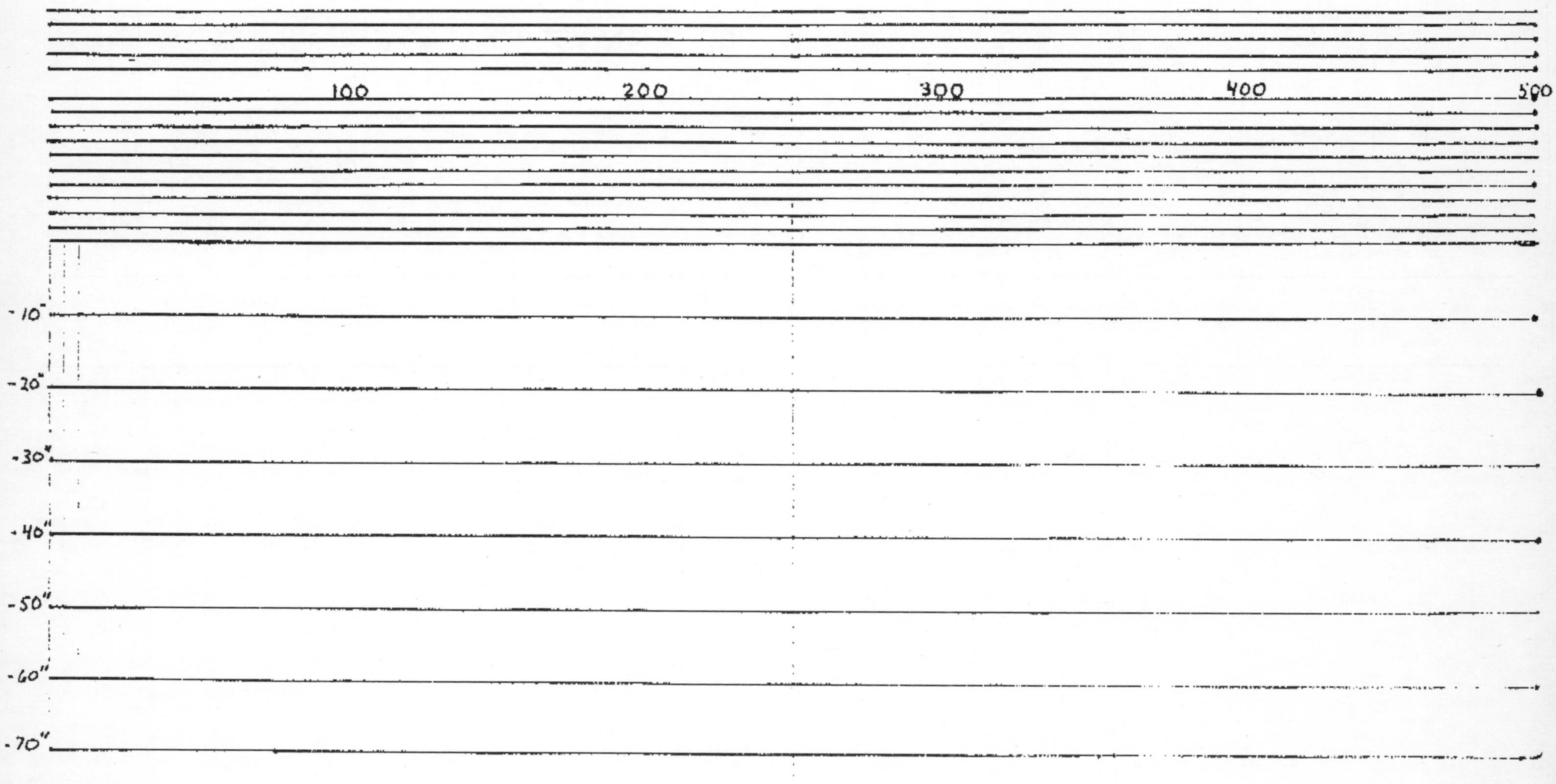

100
200
300
400
500

-10"
-20"
-30"
-40"
-50"
-60"
-70"

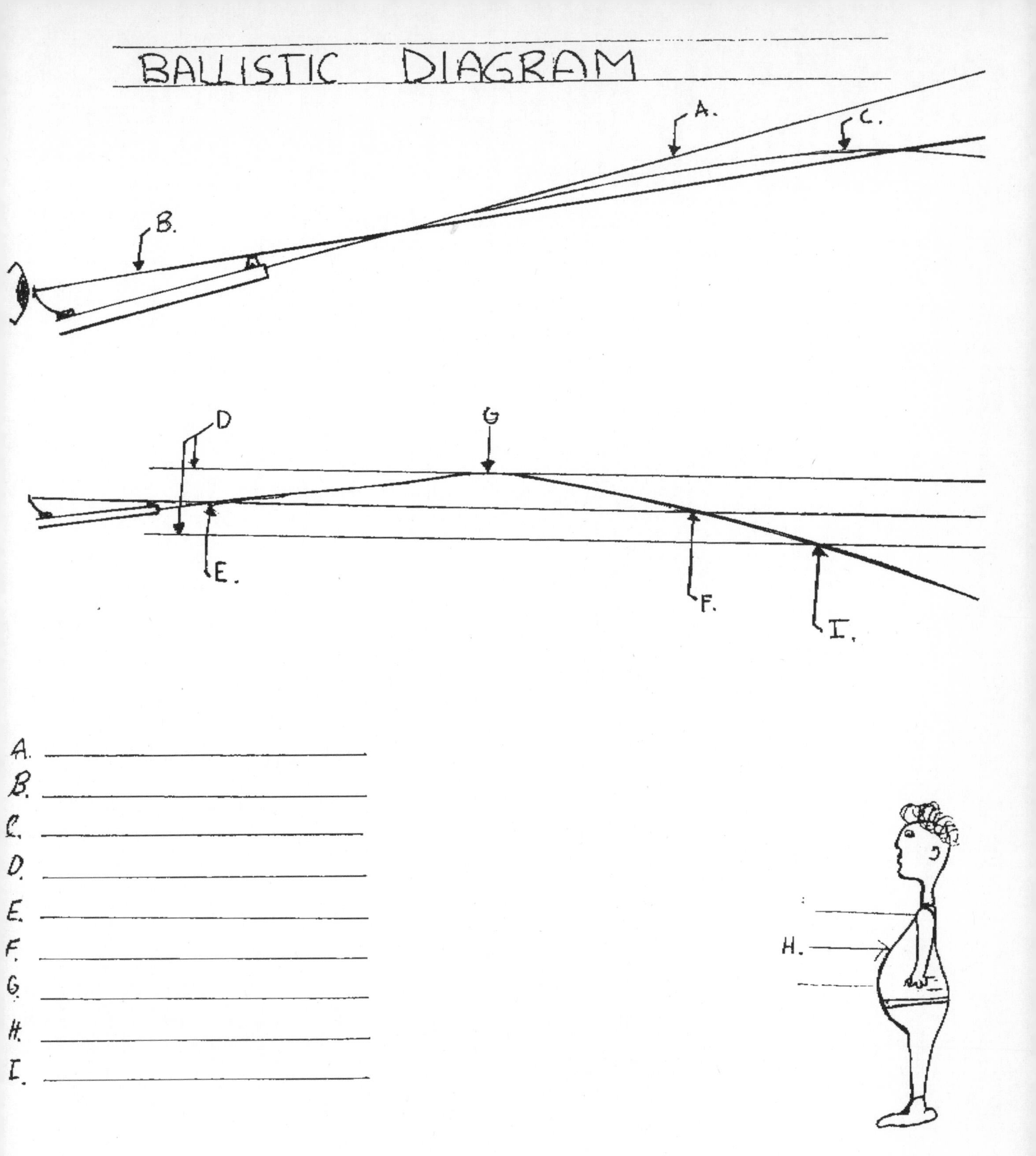
BALLISTIC DIAGRAM
A.
B.
C.
D
G
E.
F.
I.
H.
A.
B.
C.
D.
E.
F.
G.
H.
I.

LESSON OUTLINE

SPECIAL OPERATIONS

INTRODUCTION

1. Gain Attention. The continuation of a sniper program in Special Warfare will depend on how well, you as snipers can support the Seal Platoons to carry out their assigned mission or independent sniper operations.

2. Purpose. This lesson outline is constructed to better advise the Seal Scout Sniper to employ himself by giving a better understanding of weapons, tactics, and employment involving, Haskins 50 Cal S.W.S., Ship Boarding, Across the Beact operations, HAHO Airborne Insertions.

3. Training Objectives. At the end of this period of instruction you should be able to:

a. Describe the tactical employment of snipers in, 50 Cal. S.W.S. Operations, Ship Boarding, Across the Beach and HAHO Operations, involving snipers in the support of Seal Platoons, or conducting independent Seal Sniper Operations.

BODY

1. General. Before we can start, we first must know what exactly is meant by the term "Special Operation". Special Operations involving Seal Scout Snipers can be considered any mission that is not normally encountered by your standard Seal Sniper.

1. 50 Cal S.W.S

(1) Capabilities

(a) Maximum effective range - 2000 - 3000 meters

(b) Less than one minute of accuracy.

(c) 20X Leapole Vitra Ml Telescopic sight allows for:

Improved target identification and location.

(d) Used primarily in engaging hard targets

(2) Limitations

(a) Single shot bolt acitons; not capable of-semi-automatic fire

(b) Weight - 40 lbs.

(c) Requires special maintenance and repair

(3) Ammunition

(a) Armor Piercing

(b) Ball ammo.

(c) Tracer

(d) Incenerary

(e) RDX

4. Sniper Employment

A. General

(1) Effective sniping by well-trained and well-organized snipers will do more than inflict casualties and inconvenience to the enemy. It will have a marked effect on the security and morale of enemy personnel.

(2) The method by which Seal Snipers are employed will be controlled by many factors:

- Nature of the ground (Terrain)
- Distance (To the target)
- Existence of obstacles
- Insertion and extraction methods
- Degree of initiative shown by the enemy
- Type target (Hard/Soft)
- Number of snipers available.

(3) The 50 Cal S.W.S. was designed to engage hard targets from extended ranges up to 2000 meters, due to the excessive weight of this weapon, makes the employment of a two man Sniper Team impractical. The three man Sniper Team concept has been developed, to adjust for the weight problems encountered when using this weapon.

a. PT Man - will carry all communications equipment, normally will also be O.I.C.

b. Sniper - will be responsible for carrying the upper receiver of the 50 Cal S.W.S.

c. Observer - Responsibilities will be to call wind for the sniper, carry the lower receiver, bolt, muzzle break, 50 Cal ammo., spotting scope, range finder.

d. Depending on the mission and the number of snipers required to carry out the assigned mission, it may be necessary to employ more than one 3 man Sniper Team.

2. Ship Boarding Sniper Operations

A. General. The employment of snipers to support a ship boarding assault is an asset available to the Assault Force Unit Commander. The employed seal sniper team's objective is to set secutiry for the assault teams moving to the ship, while boarding, moving to their set point, during the assault and after the assault has taken place.

B. Employment. The employment of sniper teams will depend on:

- Number of sniper available
- Terrain
- Obstacles
- Number of Targets
- Distance.

The sniper team will consist of a two man sniper team (sniper/observer). These snipers should set 360° security around the target ship if possible. These sniper teams when possible should be employed in advance of the main assault force. By placing these sniper teams into position in advance, reliable information can be passed back to the main assault force of the situation on and around the target area. This information will be relayed back to the assault force unit commander to assist him in planning his operation.

(a) When using snipers in reduced light conditions all friendly forces should be marked in a way which clearly identifies them as such. Depending on what night vision devices are being used by the snipers (active or non-active I.R. sources) will depend on the marking procedures.

(b) Active I.R. Sources

1) I.R. lazer designators - I.R. tape (glint) should be worn by all friendlies.

(c) Non Active I.R. Sources

1) Night Vision Scopes - I.R. Chem lights should be worn by all friendlies.

d) Helo Support - One method available to a sniper team when employment of sniper teams in advance is not feasable.

1) Employment - Two Helos will normally be used, one port and one starboard of the target ship. Two snipers will be employed with each Helo. One sniper on each side of the aircraft.

2) The responsibilities of the aircraft is to set 360° security on the target, while the main assault force assaults the target ship.

3) Equipment.

- Night vision capability

3. Across the Beach Operations

General. Across the Beach operations involving snipers will be conducted in two methods:

- In direct support to a larger assault force

- independent sniper operations

1) Employment. (Support of Larger assault force)

A) The employment of sniper teams to support a larger assault force is an asset that can be used by the supported Unit Commander. Snipers should be employed if possible ahead of the main force (24-48 hours) to set security, and to mark the exact locations for the main force to come across the beach. Sniper teams should set security 180° to the front of the main force, 2, two man sniper teams should be used if possible.

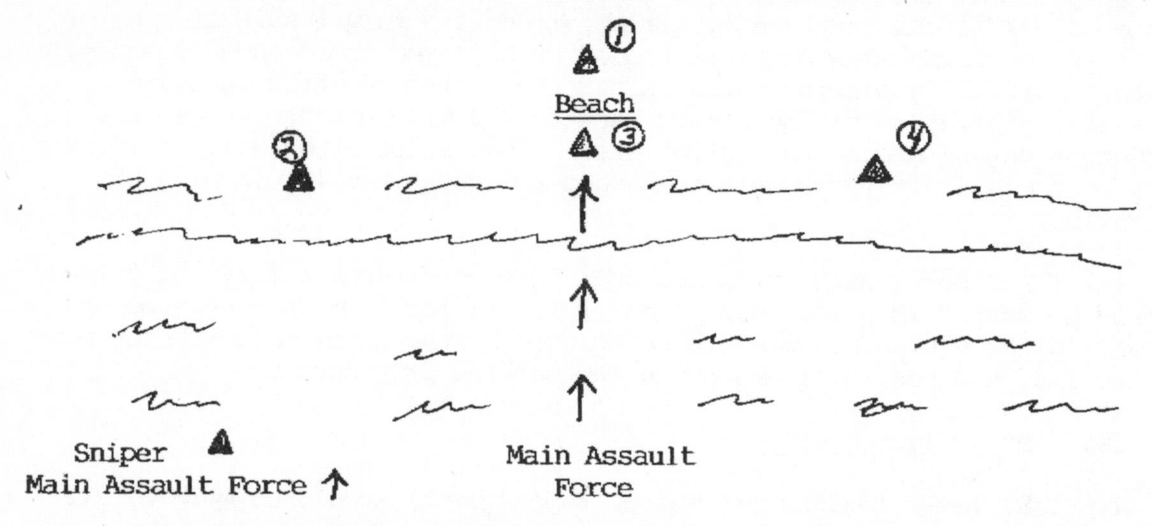

2) Snipers Responsibilities. The number one man is responsible for frontal security. Number two and four men are responsible for flank security, number three man's responsibilities are to mark the beach, link up with the main force to pass any information, and to guide the main force to the objective. After the main force has moved from their ~~extraction~~ point, sniper 1, 2, and 4 will have two options:

- They can move with the main force
  Keep flank and frontal security

- Remain behind the main force to set security, and
  help guide the main force to their extraction rally point.

If sniper teams cannot be employed ahead of the main assault force, sniper teams should come across the beach with the main assault force. Once the sniper teams are on the beach, their primary job is to set 360° security for the main assault force. Situation will dictate how snipers will be employed. Main points to remember, snipers are best employed where the main force is most vunerable.

- Moving across the beach
- Moving from the objective rally point to the object
- Moving from the objective TO THE EXTRACTION RALLY PT.
- Moving from the extraction rally point, for extraction back out to sea.

B) Independent Sniper Operations

When conducting across the beach operations that involve snipers or very small units (1-4 men) the following tactics are best used:

1. Movement Across the Beach. Snipers should swim in a swimmer pool while swimming to the beach. Once the sniper team or teams have reached a point approximately 200 meter out from the shore line should get on line facinc the shore line in order to observe the beach for any movement.

Once the sniper team(s) have reached a point where they can stand up and remove their fins the following actions should be observed:

- Before removing their fins, the sniper team should observe the beach for any enemy movement.

- Once the team is confident the beach is secure. One member at a time will remove his fins and prep any equipment for movement across the beach.

Once ready, all members will move together on line across the beach. Once across, the sniper team will proceed to their initial rally point, to prep any weapons or equipment for movement to the objective rally point.

2. Tactics. Involving movement across the beach should consider the following:

- Movement across the beach involving 1-4 men should not employ swimmer scouts. Due to the small numbers this tactic is not feasible.

- Movement from the surf zone to the beach should be carried out with all personnel on line. This tactic, will bring all guns to bear if the sniper team is compromised while moving across the beach.

4. HAHO, HALO Airborne Operations.

General.

HAHO/HALO Operations involving sniper teams can be carried out in support of a main assault force or independently.

1. In support of a Assault Force. Sniper teams (if feasible) should be employed ahead of the main force (24-48 hours) to support the main forces Unit Commander in the following areas:

- To gather information on the objective area.

- To set security for the main force's D.Z.

- Mark D.Z.

- Guide main force to the objective.

2. Independent Sniper Operations. (HAHO)

General. When conducting HAHO operations involving (1-4) sniper groupings the most important factor to consider.

The use of HAHO operations is feasible only when the need to offset the release point due to:

- Risk of compromising the D.Z.

- Threats to aircraft (AAA sights).

- Insertion aircraft cannot deviate from normal flight path.

A) Grouping. Due to the small numbers involving snipers (1-4) grouping of all personnel is of great importance. In the employment of snipers the loss of one member will most likely result in the cancellation of the assigned mission.

B) Flight Formation. The staggard stack formation is best employed (25 up/25 back) This offers the most control.

C) Base Leg. The base leg of the flight formation into the DZ should consider the following:

- Altitude. (1500 foot max) The higher the altitude the higher the risk of compromising the D.Z.

- Boxing. Right hand turns into the D.Z. Should be maintained at all times.

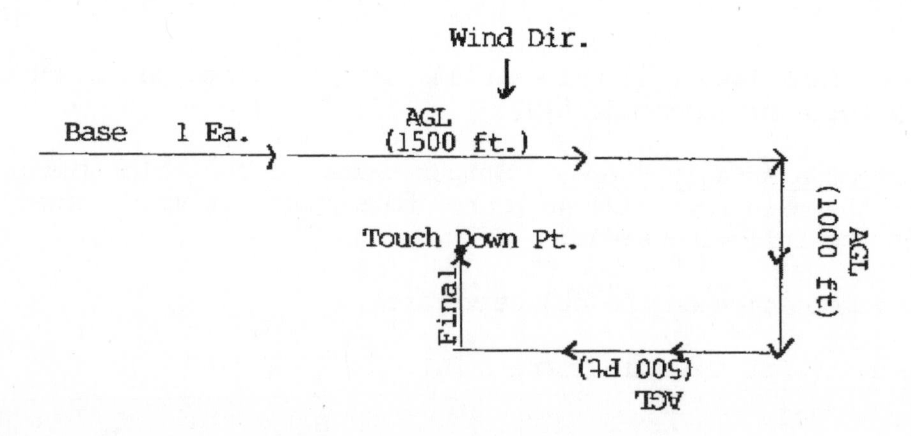

- Note: If altitude should be bled off approximately one mile out from the D.Z. The flight leader should do Flate S turns. He should give a visual signal to the rest of the formation by moving his leg in the direction of the turn.

- Compass Heading - Once exiting the aircraft and jumpers have ensured a clear airspace, they should immediately assume their compass heading into the D.Z. After the formation is heading on the correct compass heading the jumpers should assume their slots in the formation.

- Stick leader. Is usually the heaviest or most experienced jumper. His position upon exit from the aircraft will depend on what type of release the jumpmaster has selected.

Cross Wind Release - Middle of the stick

Upwind Release - Front of stick

The stick leaders responsibility is to safely navigate the flight formation to the D.Z.

- Action on the D.Z.- One word will sum this up "security".

NOTES:

1. Marking jumpers (stick leader)
   - A. Stick leader should be marked with chemical lights and strobe light
   - B. Chem lights should be attached to the backpack of the MTIX reserve. Two blue, one red chem light attached to the rear right foot and green chem light to the left foot.
   - C. This marking procedure is used to positively ID the stick leader and to aid the stick leader in signalling his intentions when making right or left turns.
   - D. The stick leader should move the leg that corresponds to this desired turn.

2. Actions on the DZ
   - A. Rally points (primary and secondary) should be predesignated prior to insertion.
   - B. Once safely on the DZ, and depending on the type of mission and situations, parachutes should be gathered up (para bags should be reused), and all jumpers should move to a central location. Once there, 360° security should be established and all chute and equipment should be buried. If time and mission dictate that chutes and supporting equipment cannot be buried, all equipment should be centralized and cammied as best possible.
   - C. Security is a prime consideration 360°. One man digs while others keep security.

LESSON OUTLINE

Aircraft Surveillance and take down

1. Purpose. To enable the SEAL Scout Sniper to employ himself during surveillance and aircraft take down.

2. Training objectives. Discribe and demostrate the tactical employment of SEAL Scout Snipers during an aircraft surveillance and takedown.

BODY

1. General. Snipers will be utilized in the intial pre-assualt surveillence to help coordinate a well planned assualt. once the assult order has been given the SEAL sniper's primary responsibilities will be security of the assualt force, surgicallly remove as many identified targets as possible and the disablement of the target aircraft, additional resposibilities:

- Gathering of pertinent information for intelligences purposes.

- Selective target elimination.

- Aircraft takedown in conjuction with breachers and assualt teams.

- Assualt force security while moving to their aircraft set point, during the assualt and after the assualt force has assualted the aircraft.

- Photographic reconn.

2. Employment. The employed SEAL snipers will be divided into two seperate elements with additional command and control element. Once all sniper elements are in place communication will be establish with the command and control element. Each element will be designated their own radio freqencey. The two elements will be designated PAPA and SIERRA (Papa is designated the port side of the aircraft and Sierra the right side of the aircraft) The number of Sniper teams to be employed will depend on the following:

- Terrain.

- Type of aircraft.

- Number of snipers availible.

- Duration of mission.

Individuals or sniper teams in each element will be numberered, begining at the front of the aircraft cockpit area (i.e. PAPA 1,3,5,7 & SIERRA 2,4,6,8,). SIERRAs will be even numbers and PAPAs will be odd numbers. Sniper or sniper teams will set security 360 degrees around the target aircraft.

3. Pre-deployment checks. The following checks should be accomplished prior to employing snipers into the field:

- Communications checks between individuals or sniper teams, command and control element, assualt team forces and ground force unit commander.

- Test fire weapons.

- Coordination between assualt teams leaders and breachers.

- Test all obsevation and photographic equipment.

- Brief all snipers on mission and individual responsibilities.

4. Resposibilities of employed sniper teams.

- Report status of all aircraft entry points. (i.e. open, close, boarding ladders, booby traps, tampering, door blockage).

- Report status of all windows, open or shut.

- Report movement by windows or entry points and possible identification of personnel (good , bad).

- Report areas of freqent use and identify by whom.

- Photograph any personnel if possible.

- Report any weapons and what type if possible.

- Describe any personnel and clothing they are wearing, physical characteristics of any personnel.

- Report any siduation or event that could compromize the assualt force or any possible threat to any hostages.

- During the assualt stage report statis of targets to command and control element (the word "green" or "red", green meaning sniper has identified a threat and can engage that threat, red meaning sniper has no target to engage.)

- During the assualt teams movement to the aircraft the words ("stop & go") will be relayed to the command and control element which will be relayed to the assualt force's team leader. The word "stop" meaning possible compromize stop all movement, "go" meaning continue with movement. (This is normaly done when the assualt force is making their move to the aircraft set point.)

4. Emergency assualt. If a situation developes which an emergency assualt (normaly conducted if mission is compromized or extream threat to hostages exist) is necessary, snipers will remain in postion or if sniper have not be able to reach their (F.F.P.) final firing position that will take up an hasty firing postions. Thier responsibilities will be to engage any threats to the assualt team and hostages as they appear untill the assualt has commented their assualt on the target aircraft. After the assualt has taken place all snipers will remain in place in order to maintain 360 degree security of the assualt force unless so directed by the command and control element. (This is normaly done for further identification of hostagaes and terrorists. A minimal sniper force should be left in postion to maintain security).

a. The use of cover fire or diversionary fire through the upper 1/3 of the aircraft to encourage all personnel inside the aircraft to get down untill the assault team arrives.

5. Deliberate assault. This is when all sniper and assualt teams have the time to fully prepare their selfs for the assualt.The snipers will preform the following tasks:

- Move to (F.F.P.) finial firing postion and establish postion by constructing range card, setup observation and photographic equipment.

- Comence surveillance.

- Depending on sniper's postion and area of responsibility, select proper bullet. (Solid lead or Sabo round for aircraft windshield copper jaket ammo should not be used because of the frag hazard copper jacket ammo can be used to engage targets in open doors or windows.)

- Test communiction with all units. There should be three sperate radio freqs:

1. Primary - Sniper to Sniper command and control element.

2. Secondary - Sniper to assult team leader.

3. Secondary - Sniper to sniper.

NOTE: When establishing communication all snipers should have an capiblity to engage targets at one time in unison. This is referred to as synchronized or synchro firing. BY engaging all targets at one time there is no time for the threats to react to an siduation where one of their commrads has taken a hit, which gives the bad guys time to kill an hostage.

Once communication has been established between snipers and command and control and the siduation warrents to engage any threats by synchro fire (normally this is done in unison with breaching the aircraft just slighly before the breach. This will reduce the number of bad guys the assualt force has to deal with.) the communication format is as follows:

- Snipers stand by "GREEN".

- Snipers "READY", "READY", "READY" , "FIRE".

all targets will be engaged, if the word fire is not said sniper will not engage target, if there is a delay in the sequence the format will be completely started over.

The word "BLACK" means donot engage targets donot fire.

4. Set security for breaching and assualt teams.

5. Upon aquistion of correct number of targets (ground force commander will relay sniper command and control to engage targets, command and control will only relay the order to shoot upon the order of the ground force commander.)

6. Essential equipment.

- Remington model 700. or 300 windmag.

- Communictions equipment MX-300R OR MX-360. hand raidos.

- Spotting scope.

- Binos.

- Map of area.

- Log book if working in sniper teams.

- Range card.

- Range finder if avaible.

- Radio pouch.

- 40 rounds 7.62 Or 300 windmag.

- Standard issue side arm with 30 rounds.

- Red lens flash light.

- Strobe light.

- Depending on the siduation the option of using night vision scopes should be included with all snipers.

- Secondary weapon with M-845 night vistion device.

- Terrain and climate will dictate uniform.

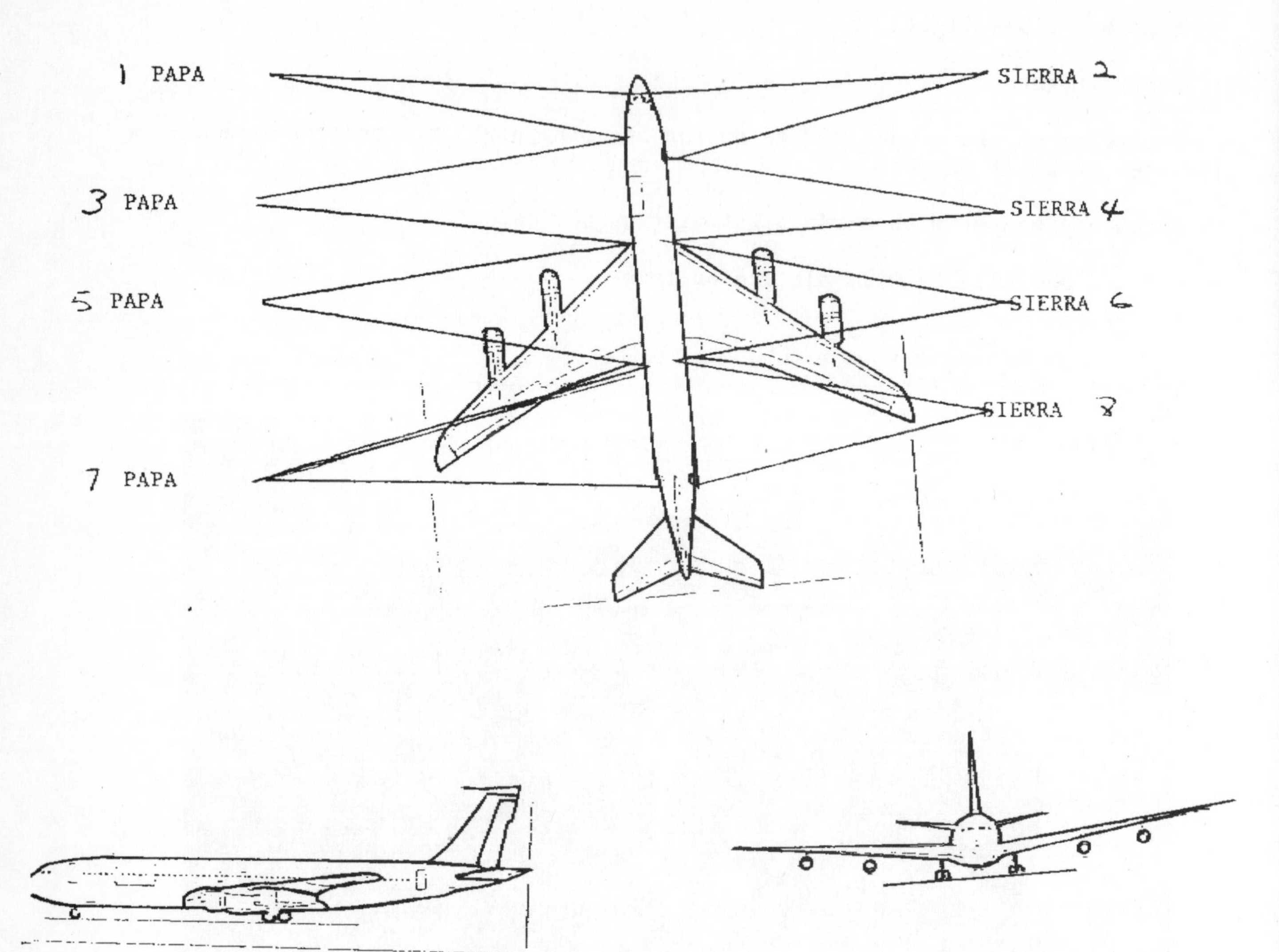

Boeing 707 320C four-turbofan passenger cargo transport aircraft. (Pilot Press)

COCKPIT AND FRONT EXIT
WINDOWS FROM FRONT EXIT TO LEADING EDGE OF WING
ALL WINDOWS OVER WING AND ALL WINDOW EXITS
WINDOWS AFT OF WING AND REAR EXIT

FIRING POSITIONS CAN BE THE SAME AS OBSERVATION POSITIONS BUT CAN BE MOVED TO COVE SUSPECT AREAS OR MORE HEAVILY USED EXITS OR WINDOWS.

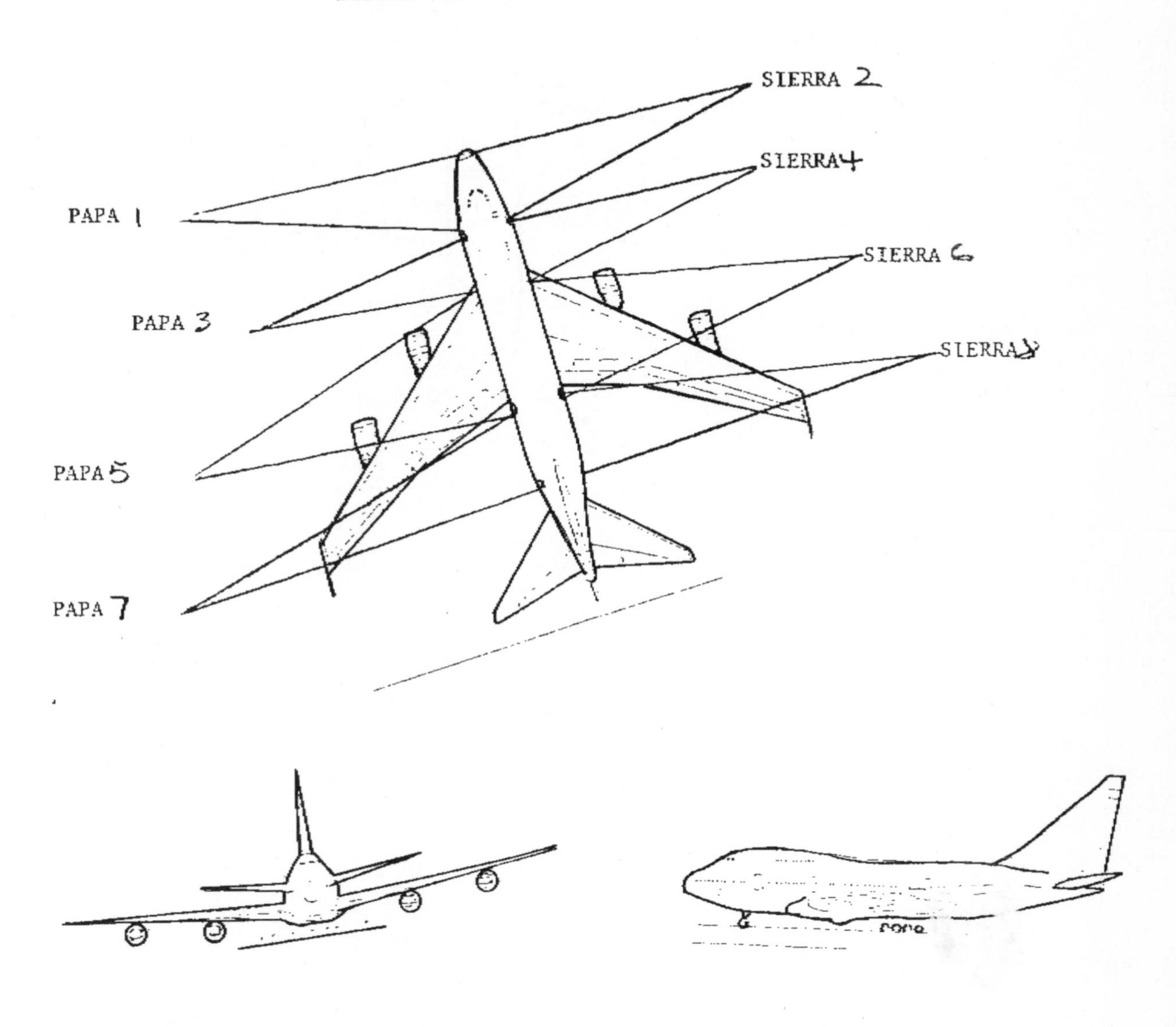

Boeing 747SP short-fuselage range version of the 747. (Pilot Press)

COCKPIT AND LOWER WINDOWS TO FRONT DOOR
FRONT DOORS TO WING EXIT AND UPPER DECKS
WING EXIT FRONT TO WING EXIT REAR
REAR WING EXIT TO REAR DOOR

FIRING POSITIONS CAN BE THE SAME AS OBSERVATION POSITIONS BUT CAN BE MOVED TO BEEF UP SUSPECT AREAS OR COVER HEAVILY USED EXITS OR WINDOWS.

BOEING 737-200

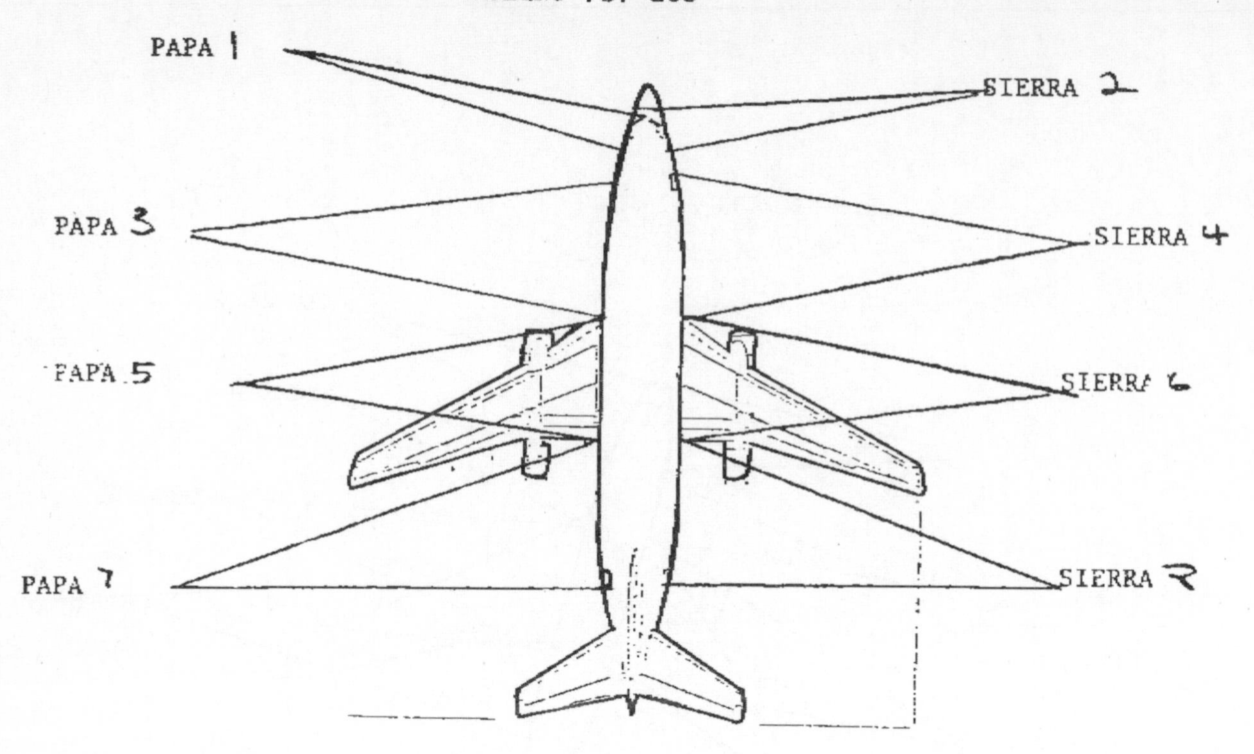

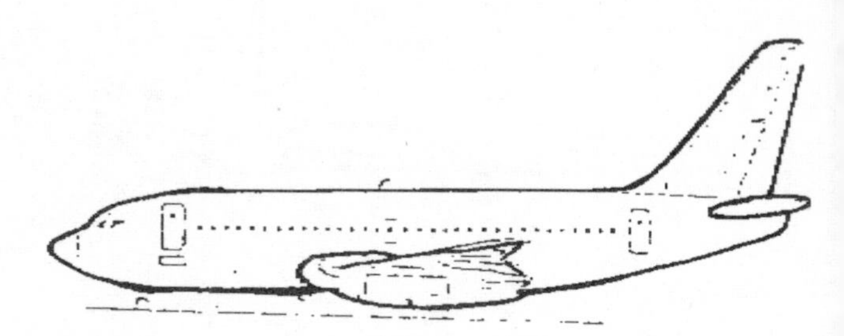

Boeing 737-200 twin-turbofan, short-range transport. (Pilot Press)

COCKPIT
FRONT EXIT AND WINDOWS TO WING
WING WINDOWS AND WING EXITS
AFT OF WING WINDOWS AND REAR EXIT

FIRING POSITIONS CAN BE THE SAME AS OBSERVATION POSITIONS BUT CAN BE MOVE TO BEEF UP MORE SUSPECT AREAS OR TO COVER HEAVILY USED DOOR OR WINDOWS.

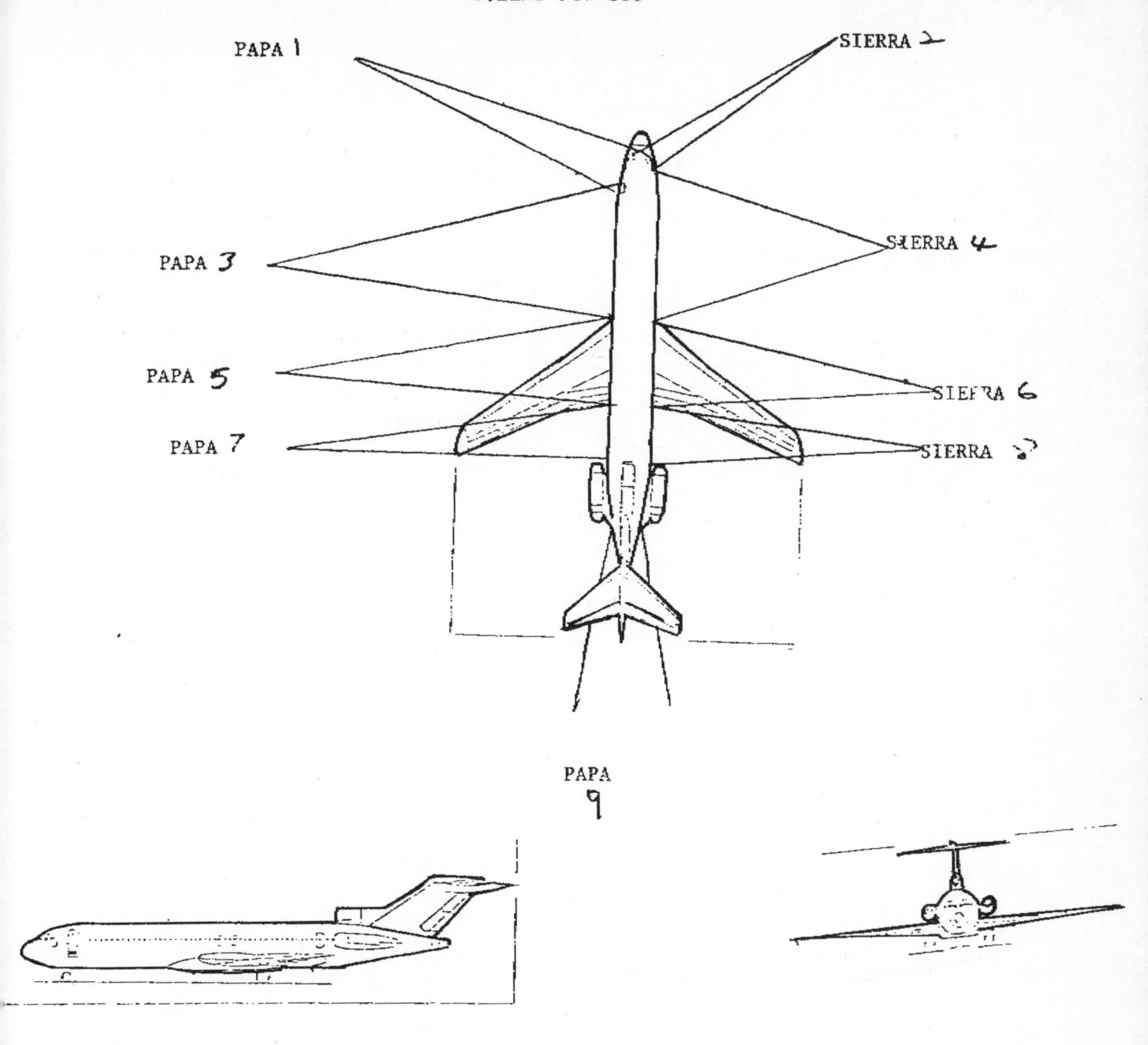

Boeing 727-200 three-turbofan, short/medium-range transport. (Pilot Press)

PAPA COCKPIT AND FRONT EXIT
SIERRA COCKPIT
PAPA WINDOWS TO WING
SIFRRA WINDOWS FROM COCKPIT TO LEADING EDGE OF WING
WING EXITS AND WINDOWS OVER WINGS
WINDOWS AFT OF WINGS AND REAR DOOR
APA REAR RAMP

FIRING POSITIONS CAN BE THE SAME AS OBSERVATION POSITIONS BUT CAN BE MOVED TO MORE SUSPECT AREAS OR TO COVER HEAVILY USED WINDOWS OR EXITS.

BOEING 767-200

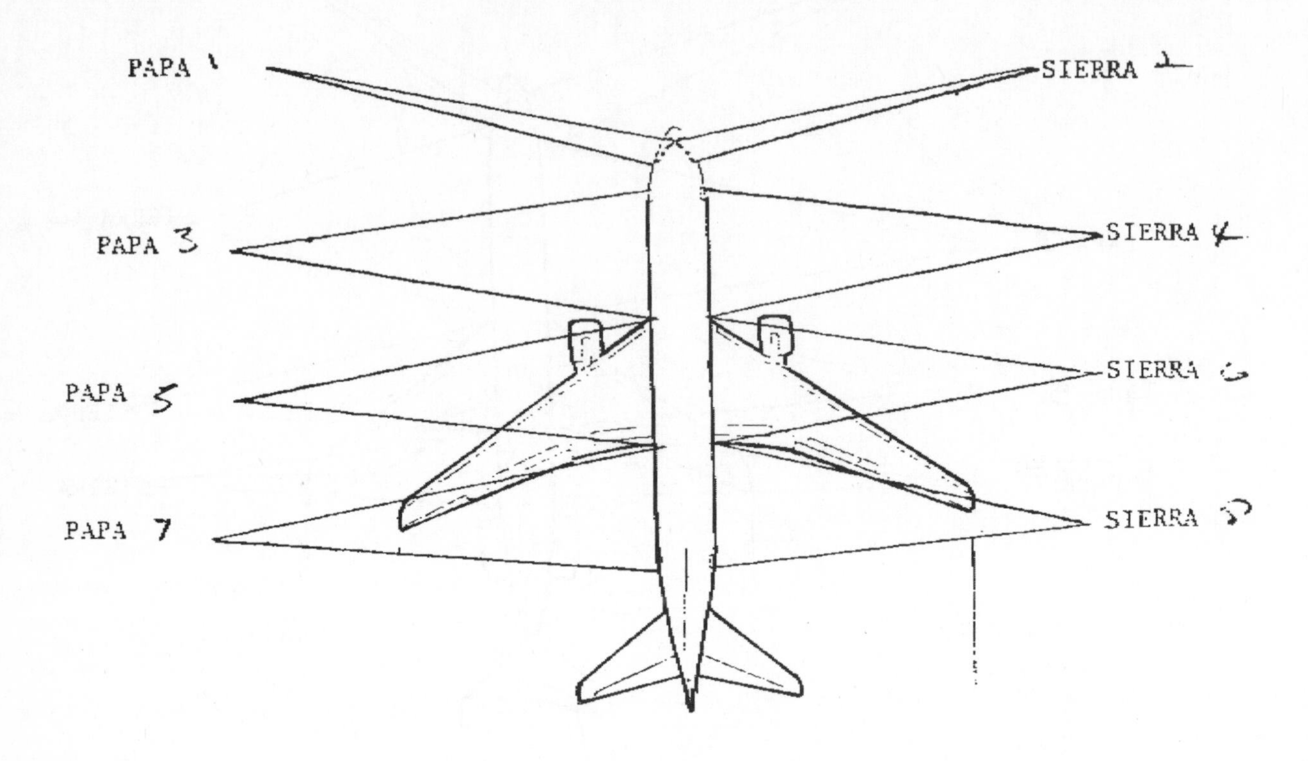

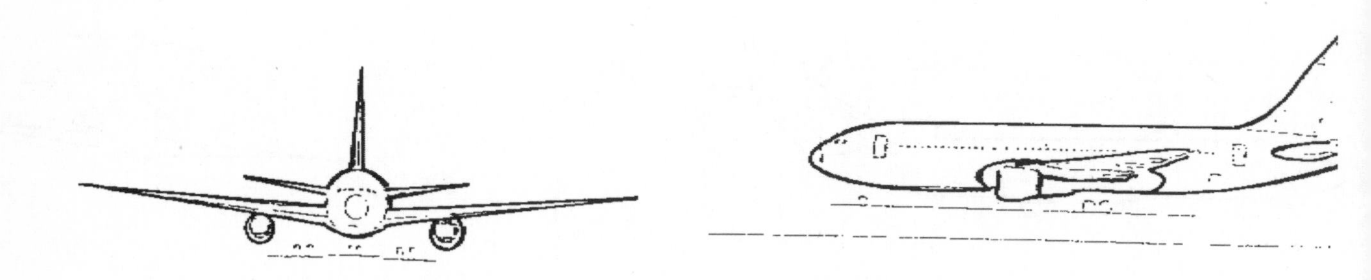

Boeing model 757-200 wide-bodied, medium-range, commercial transport aircraft. (Pilot Press)

COCKPIT

FRONT EXIT AND WINDOWS TO WING LEADING EDGE

ALL WINDOWS OVER WINGS AND WING EXITS

WINDOWS AFT OF WING AND REAR EXIT

FIRING POSTIONS CAN BE THE SAME AS OBSERVATION POSITONS BUT CAN BE MOVED TO COVER SUSPÈCT AREAS OR HEAVILY USED EXITS OR WINDOWS.

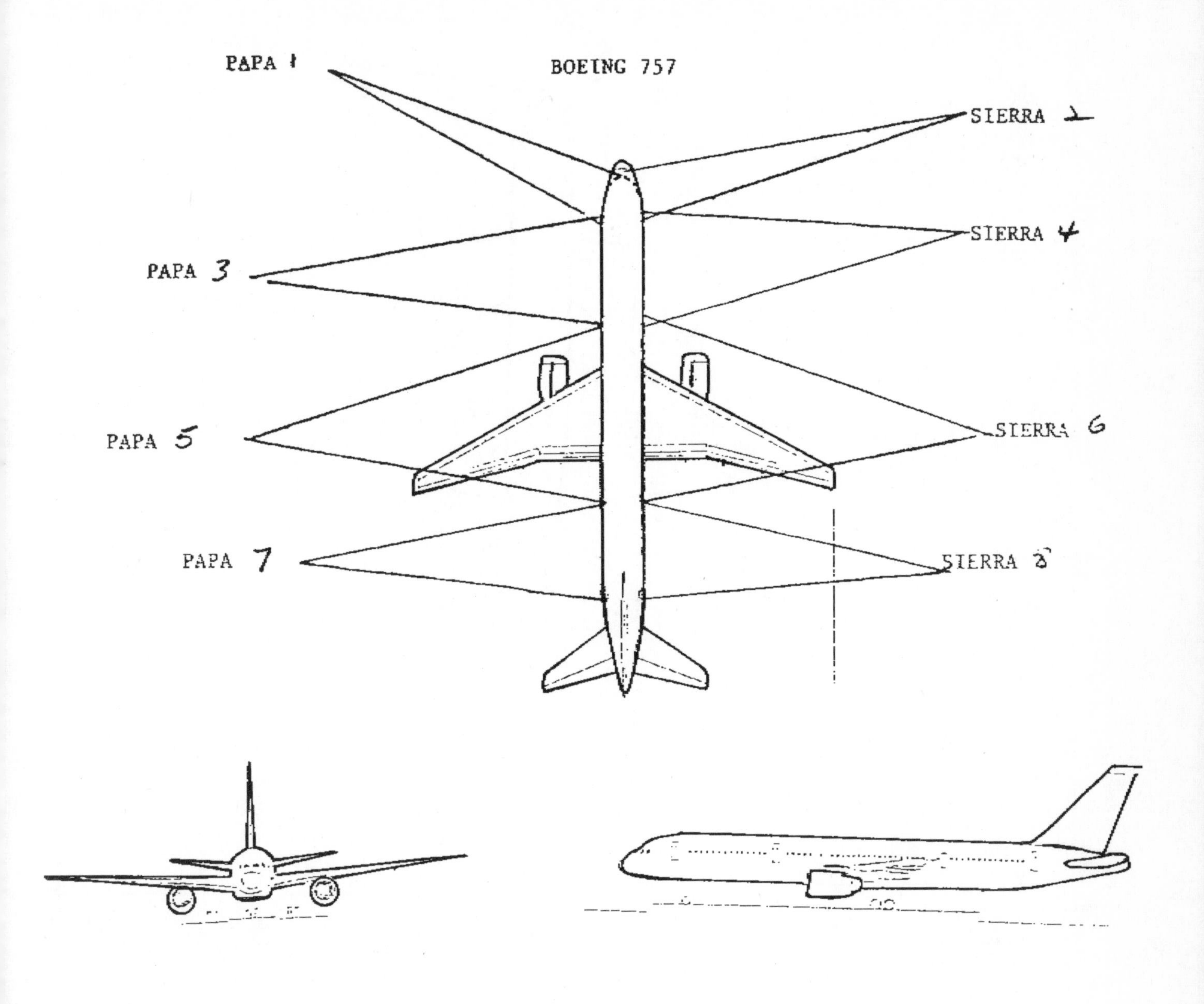

Boeing model 757 twin-turbofan, medium-range, commercial transport aircraft. (Pilot Press)

COCKPIT AND FRONT WINDOW
WINDOWS TO SECOND EXIT DOOR AND DOOR
WINDOWS TO THIRD EXIT DOOR AND DOOR
WINDOWS TO REAR EXIT DOOR AND DOOR

ING POSITIONS CAN BE THE SAME AS OBSERVATION POSITIONS BUT CAN BE CHANGED TO COVER PECT AREAS OR MORE HEAVILY USED EXITS AND WINDOWS.

MCDONNELL DOUGLAS
DC-10

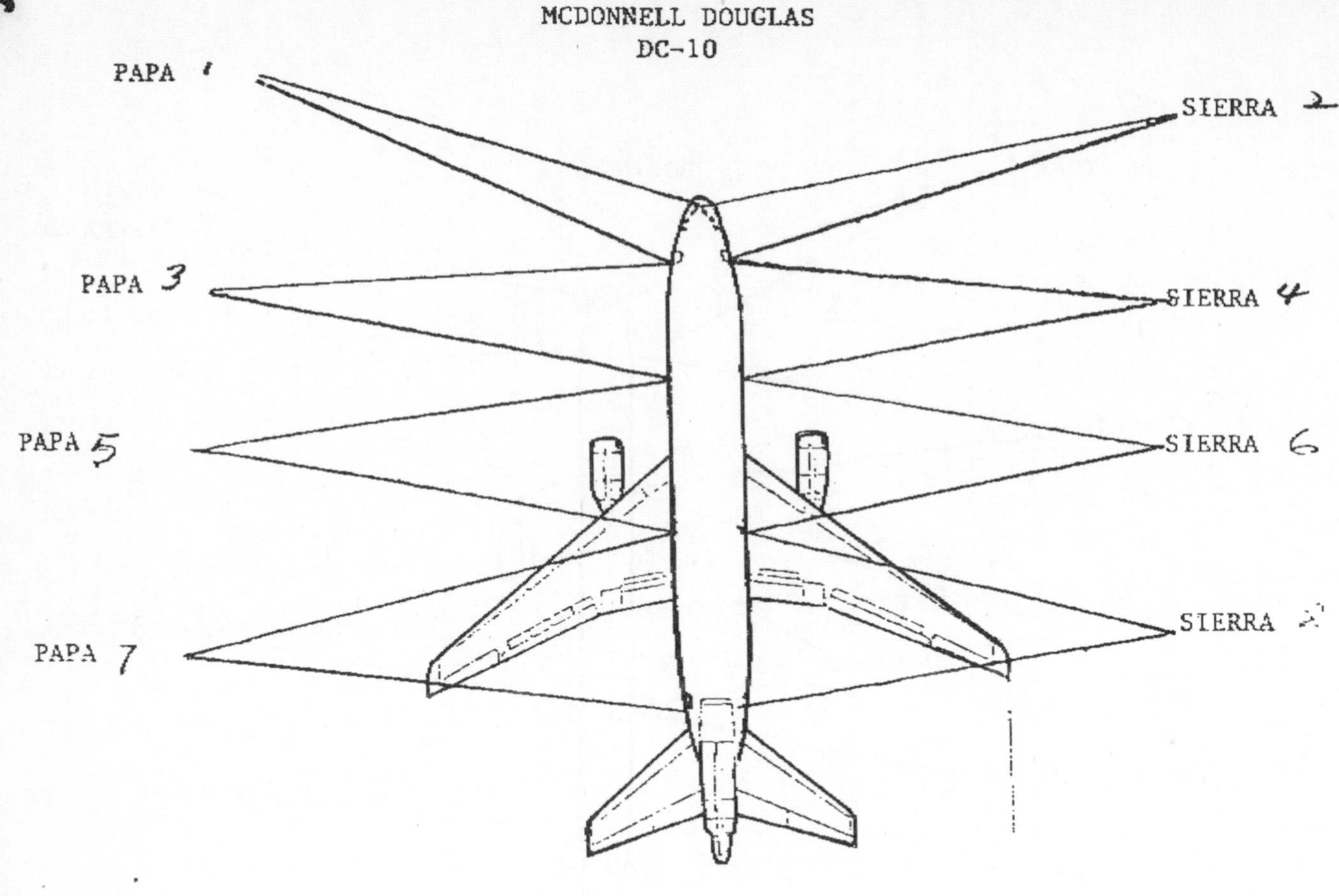

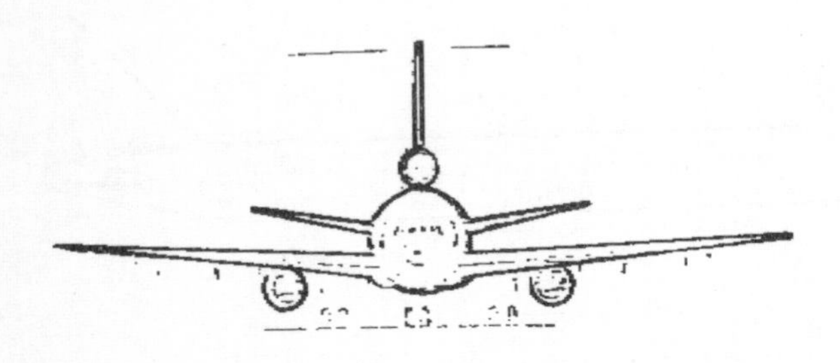

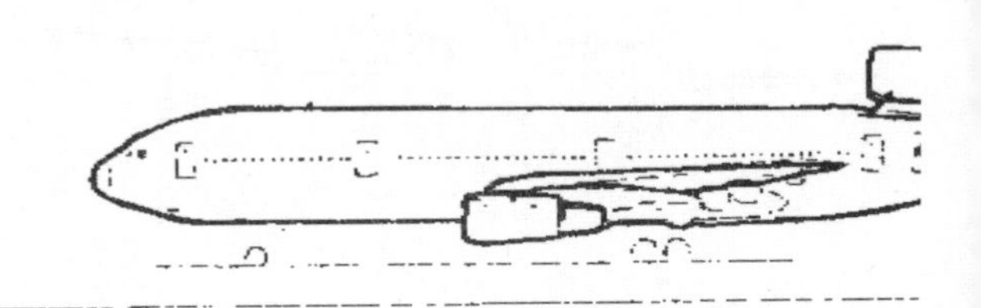

McDonnell Douglas 30 high-capacity, three-engined transport. (Pilot Press)

PAPA COCKPIT AND FRONT EXIT
PAPA SECOND EXIT AND ALL WINDOWS FORWARD TO FRONT EXIT
PAPA WING EXITS AND WINDOWS TO SECOND EXIT
PAPA REAR EXIT AND WINDOWS TO WING EXIT

FIRING POSITIONS CAN BE THE SAME AS THE OBSERVATION POSITIONS BUT CAN BE MOVED TO BEEF UP MORE SUSPECT AREAS OR HEAVILY USED EXITS OR WINDOWS.

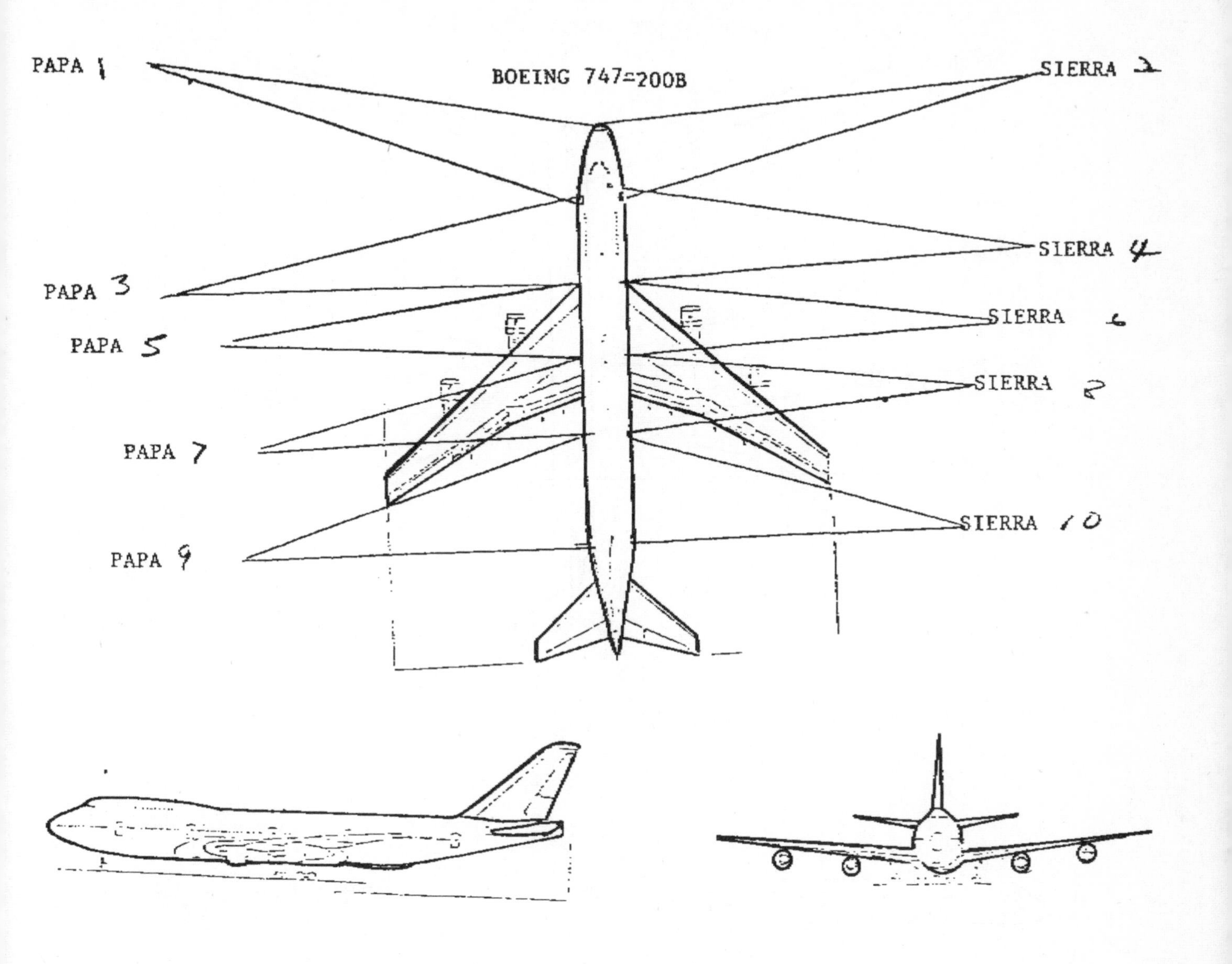

Boeing 747-200E four-turbofan, heavy-transport aircraft. (Pilot Press)

COCKPIT AND LOWER DECK TO FRONT DOOR
FRONT DOOR TO WING LEADING EDGE AND UPPER DECK
ALL WINDOWS ABOVE WING TO THE MIDDLE WING EXITS
ALL WINDOWS ABOVE WING FROM THE MIDDLE WING EXIT TO EXIT BEHIND WING
EXIT BEHIND WING TO REAR EXIT AND REAR EXIT

FIRING POSITIONS CAN BE THE SAME AS OBSERVATION POSITIONS BUT CAN BE MOVED TO MORE SUSPECT AREAS OR TO COVER HEAVILY USEDEXITS OR WINDOWS.

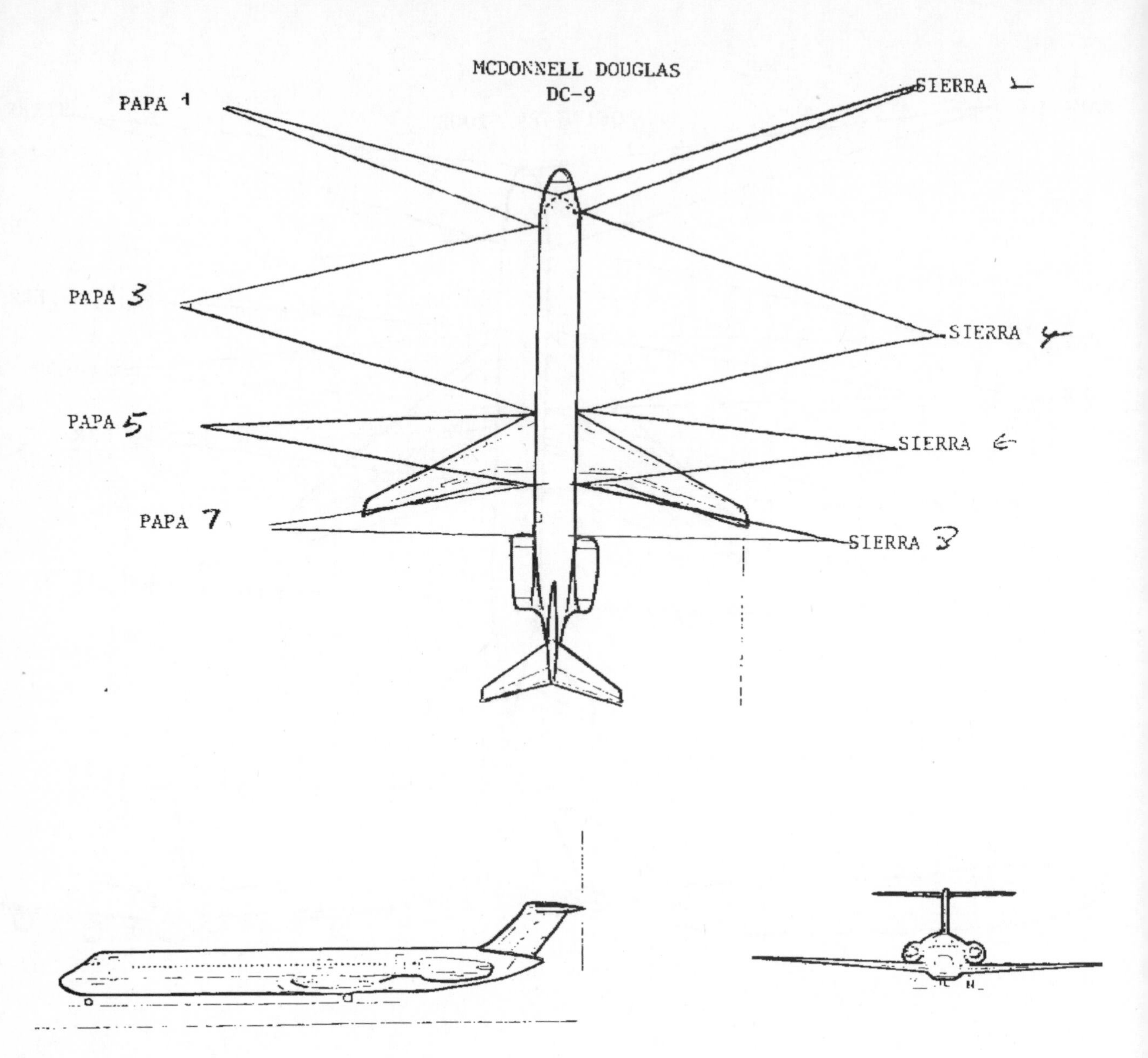

McDonnell Douglas DC-9 Super 20 "stretched" version of this twin-turbofan transport. (Pilot Press)

PAPA COCKPIT AND FRONT EXIT
SIERRA COCKPIT
WINDOWS TO FRONT OF WING
WING WINDOWS AND EXITS
WINDOWS AFT OF WING AND REAR EXIT

FIRING POSITIONS CAN BE THE SAME AS OBSERVATION POSITIONS BUT CAN BE CHANGED TO BEEF UP MORE SUSPECT AREAS OR HEAVILY USED WINDOWS OR DOORS.

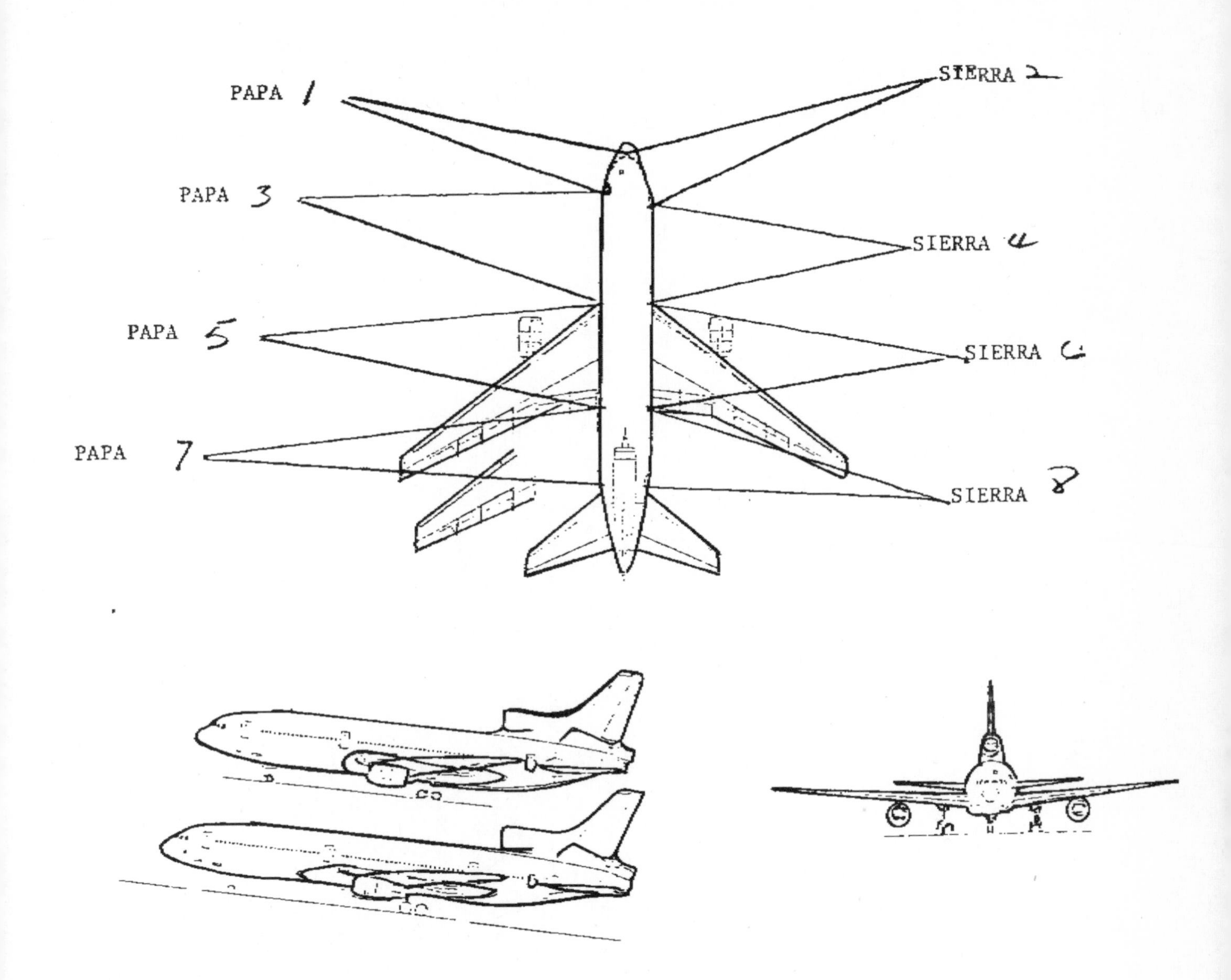

Lockheed L-1011-500 TriStar extended-range, wide-bodied transport with added side view (bottom) and scrap view of wingtip of L-1011-100. (Pilot Press)

SURVEILANCE POSITIONS AND INITIAL FIELDS OF FIRE

COCKPIT AND FRONT EXITS

ALL WINDOWS BETWEEN FRONT EXIT AND SECOND EXIT AND SECOND EXIT HATCH

ALL WINDOWS BEHIND SECOND EXIT AND OVER WINGS

WINDOWS AFT OF WING AND EITHER ONE OR TWO REAR EXITS

FIRING POSITIONS CAN BE SAME AS OBSERVATION BUT CAN BE MOVED TO BEEF UP MORE SUSPECT AREAS OR HEAVILY USED ESITS.

PICKUP ZONE AND LANDING ZONE OPERATIONS

3-1. SELECTION OF A PZ/LZ site

a. Dimensions.

(1) The landing point should be large enough to have a cleared circular area for landing of at least 25 meters for day or night. These dimensions are the minimum requirements for safety considerations. (Refer to para 3-8 for correct dimensions of cleared areas.)

(2) The landing site should have the capacity to accomodate the number, formation, and type of aircraft using the site.

b. Surface Conditions.

(1) The PZ/LZ should be free of obstacles and any loose debris that might damage the aircraft or obstruct the pilot's vision. All obstacles should be removed if possible. All obstacles that cannot be removed should be well marked. Vegetation over 18 inches in height should be cut and removed.

(2) The surface of the site should be firm enough to support the weight of the aircraft.

(3) Ground slope is determined by mathematical computation. To determine the amount of slope, the following method is used:

V = Vertical Distance
H = Horizontal Distance

V is found by computing the difference between the lowest point and the highest point on the site using the map. This number is multiplied by 57.3.

H is the length of the cleared area.

$$\frac{V \times 57.3}{H} = \text{degrees of slope.}$$

EXAMPLE: $\frac{V = 25' \times 57.3 = 1432.5}{H = 500 = 500}$ = 2.8 degrees rounded up to 3 degrees ground slope.

3-2. GROUND SLOPE REQUIREMENTS

Since landing to a sloped surface is somewhat more difficult than landing on a flat and level surface, the ground slope should be taken into consideration before beginning establishment of the site. The following rules should be

considered when the landing surface has slope. During night operations, the pilot should be advised that he will be landing on a sloped surface.

a. Avoid landing downslope.

b. Utility and observation type aircraft will not land when the slope exceeds 7 degrees.

c. Large utility and cargo type aircraft will be issued an advisory when the slope is 7 to 15 degrees.

d. Ideally, the surface of the landing point should be level.

3-3. APPROACH AND DEPARTURE DIRECTIONS

Attempt to have the aircraft approach and depart over the lowest obstacles. An obstacle ratio of 10:1 for planning purposes is used, but this may vary depending on the aviation unit's SOP. If you are unable to achieve the desired 10:1 ratio, consider whether the aircraft will be landing loaded or unloaded. If landing loaded, use the greater obstacle ratio on the approach end. If departing loaded, have the greater obstacle ratio on the departure end.

Figure 3-1. Obstacle ratio.

3-4. LANDING FORMATIONS

See figures 3-2 through 3-8

3-5. NUMBER AND TYPE OF AIRCRAFT.

a. Section: 4 UH-1H; 3UH-60; 2 CH-47

b. Platoon: 8 UH-1H; 6 UH-60; 4 CH-47

c. Company: 25 UH-1H; 19 UH-60; 12 CH-47

3-6. WIND

If the wind below 1,000 feet mean sea level (MSL) exceeds 90 degrees, and deviates in excess of 45 degrees from the long axis of the landing site, the land heading should be adjusted or an aircraft advisory issued. The allowable wind velocity decreases as the density altitude increases. Because of the design of the helicopter, if can only accept a minimum velocity of wind from certain directions.

a. Aircraft can land with a tail wind of 0 to 5 knots.

b. Aircraft can land with a cross wind of 0 to 9 knots.

c. Aircraft can land with a head wind of 10 knots and above.

3-7 DENSITY ALTITUDE

a. Density Altitude. Density altitude determines the actual lift capability of the aircraft for that particular day. There are three environmental factors which affect the performance of helicopters: altitude, temperature, and humidity. As any of these factors increases, the performance of the aircraft decreases. Optimum aircraft performance is obtained during a cool day with the field elevation as close to sea level as possible and with relatively low himidity.

b. Marking of a PZ/LZ.

(1) Helicopter landing points are categorized by size to accomodate different types of aircraft. The size of the landing point to be used is determined by the type aircraft employed, the aircraft's mission (troop lift, slingload, internal cargo), and the extent of coordination with the supporting aviation unit. The five sizes of landing points are indicated below and are applicable to both day and night operations.

Size	Cleared Area Diameter
1	25 meters
2	35 meters
3	50 meters
4	80 meters
5	100 meters

(2) The landing point size for the different aircraft is indicated below. The distance between landing points is equivalent to the diameter of the cleared area.

Type Aircraft	Landing Point Size
OH-58	1
UH-1H	2
UH-60	3
AH-1G	2
CH-47	4
CH-54	4
CH-53	4
AH-64	3

NOTE: When integrating a slingload point with a landing site, the slingload point will normally occupy the area considered to be the last landing point in the formation. If the slingload aircraft is different in type then the landing aircraft, the distance from the center of the last landing point to the center of the slingload point will be 100 meters. If the slingload aircraft is of the same type as the landing aircraft, the distance between

the center of the last landing point ahd the center of the slinghood point will be equivalent to the diameter of slingload point. In eaither case, the slingload point will be the last point in the formation. The slingload point will always be either a size 4 or size 5 landing point, regardless of type aircraft transporting slingloads.

(3) The position of the signalman is approximately 40 meters to the aircraft's right front for landing aircraft.

NOTE: These dimensions are general guidelines and may change depending on the supporting aviation unit's SOP and the extent of coordination.

9. Chalk Numbers.

HELICOPTER FORMATIONS AND CHALK NUMBERS		
Staggered Trail Left	Diamond	Staggered Trail Right
Heavy Left Formation	Vee	Heavy Right Formation
Echelon Left	Trail	Echelon Right

Figure 3-8. Formations with chalk numbers

3-9. MARKING TOUCHDOWN POINTS

a. Daylight. The number one touchdown point is marked by a signalman. However, the aviation unit SOP will normally dictate how the site is to be marked during daylight.

b. Night. The number one touchdown point is marked with an inverted Y or a landing T. The inverted Y is the most preferred method.

(1) Inverted Y. The Y is best used for an approach initiated from terrain flight altitudes. The desired touchdown point is midway between the front two lights with the fueslage of the aircraft aligned with the stem lights. A minimum of four lights is used.

INVERTED Y

Landing Direction

Stem Light

7M

Stem Light

14M

Left Leg Light

14M

Right Leg Light

7M 7M

Base Point

Figure 3-9. Inverted Y.

(2) Landing T. The T is best used for approaches initiated from air altitudes above 500 feet above ground level (AGL). The apparent distance between the lights in the stem of the T can be used as a reference for maintaining a constant approach angle. The approach should be terminated in the upper left portion of the T.

LANDING T

5M 5M

8M

Base Light

8M

Figure 3-10. Landing T.

(3) Additional touchdown points

(a) Utility aircraft. Marked with 2 Lights 5 meters apart.

(b) Cargo Aircraft. Marked with 2 lights 10 meters apart.

3-10 PLACEMENT OF VAPI

When a visual approach path indicator (VAPI)* is used, it will be placed according to the following:

*NOTE: VAPI must be checked after each departure.

(1) Inverted Y. The Y is best used for an approach initiated from terrain flight altitudes. The desired touchdown point is midway between the front two lights with the fuselage of the aircraft aligned with the stem lights. A minimum of four lights is used.

INVERTED Y

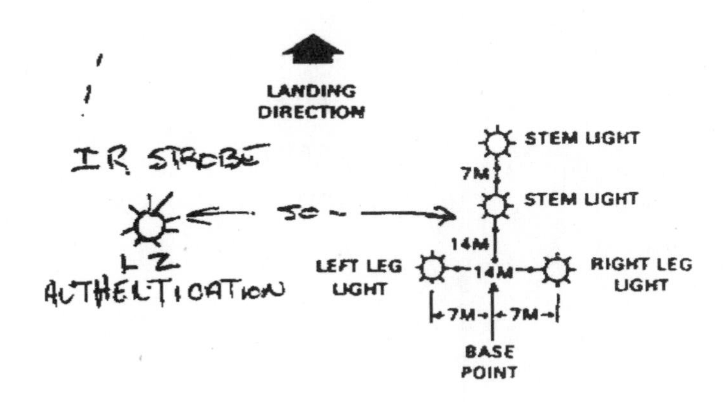

Figure 3-9. Inverted Y.

(2) Landing T. The T is best used for approaches initiated from air altitudes above 500 feet above ground level (AGL). The apparent distance between the lights in the stem of the T can be used as a reference for maintaining a constant approach angle. The approach should be terminated in the upper left portion of the T.

LANDING T

5M 5M
8M
BASE LIGHT
8M

Figure 3-10. Landing T.

(3) Additional touchdown points.

(a) Utility aircraft. Marked with 2 lights 5 meters apart.

a. Inverted Y.

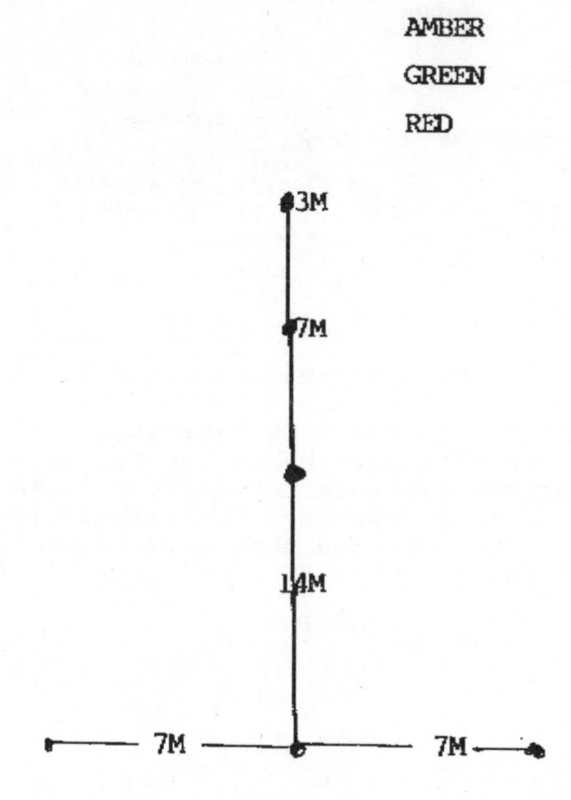

Figure 3-11. VAPI when used with Y.

b. Landing T.

AMBER

GREEN

RED

15M

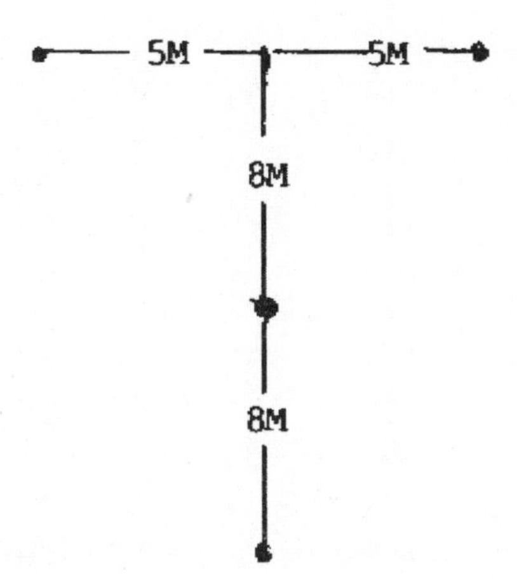

Figure 3-12. VAPI when used with T (utility aircraft).

AMBER

GREEN

RED

25M

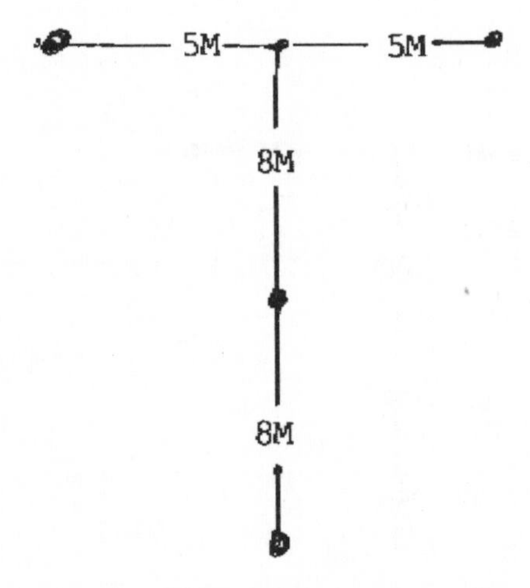

Figure 3-12.1 VAPI when used with T (cargo aircraft).

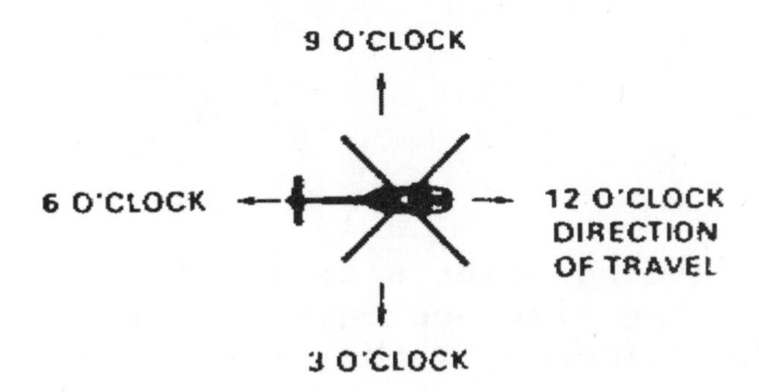

Figure 3-14. Clock method of direction.

Size	Cleared Area Diameter
1	25 meters
2	35 meters
3	50 meters
4	80 meters
5	100 meters

(2) The landing point size for the different aircraft is indicated below. The distance between landing points is equivalent to the diameter of the cleared area.

Type Aircraft	Landing Point Size
OH-58	1
UH-1H	2
UH-60	3
AH-1G	2
CH-47	4
CH-54	4
CH-53	4
AH-64	3

3-6. WIND

If the wind below 1,000 feet mean sea level (MSL) exceeds 90 degrees, and deviates in excess of 45 degrees from the long axis of the landing site, the land heading should be adjusted or an aircraft advisory issued. The allowable wind velocity decreases as the density altitude increases. Because of the design of the helicopter, it can only accept a minimum velocity of wind from certain directions.

a. Aircraft can land with a tail wind of 0 to 5 knots.

b. Aircraft can land with a cross wind of 0 to 9 knots.

c. Aircraft can land with a head wind of 10 knots and above.

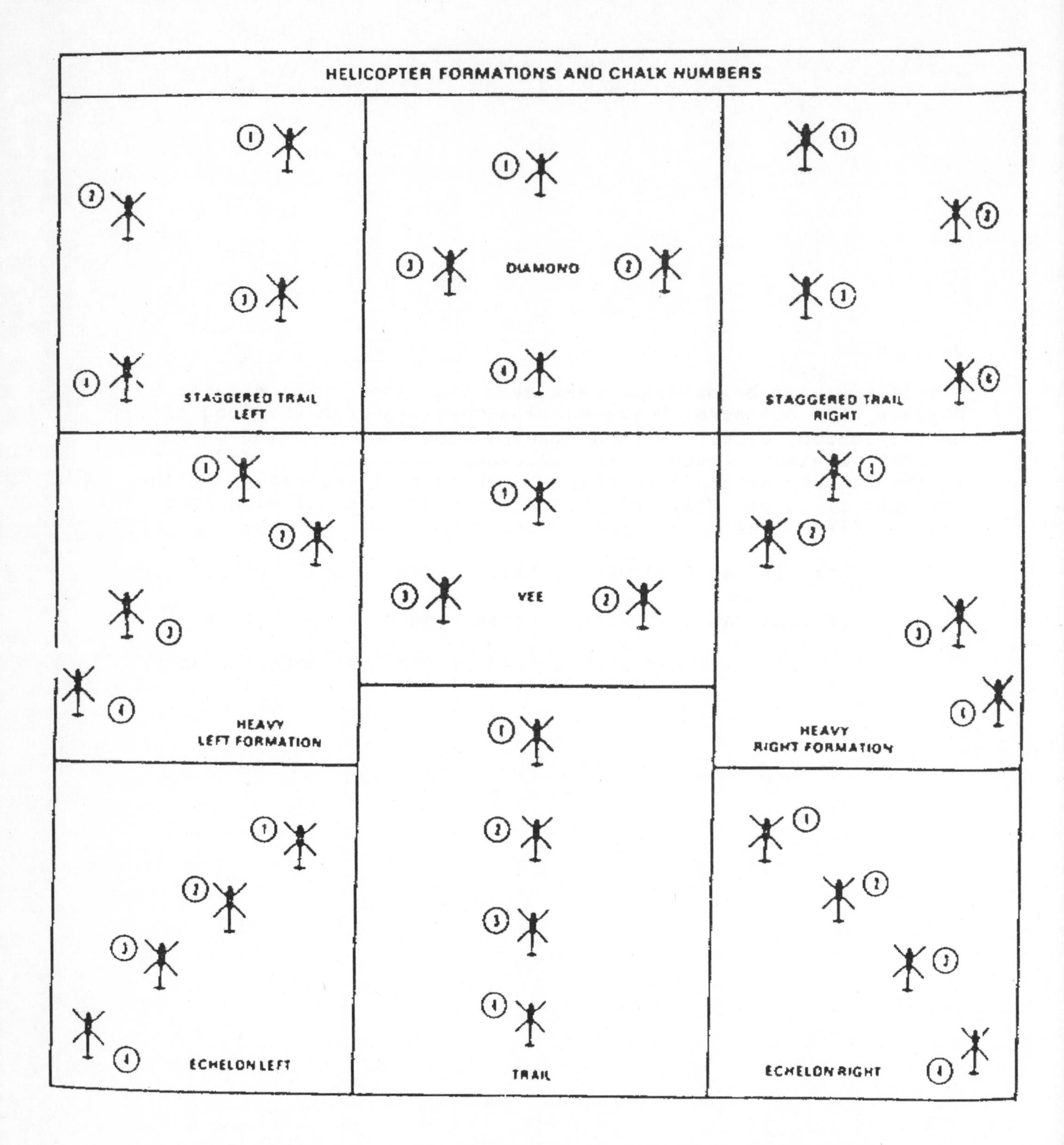

Figure 3-8. Formations with chalk numbers.

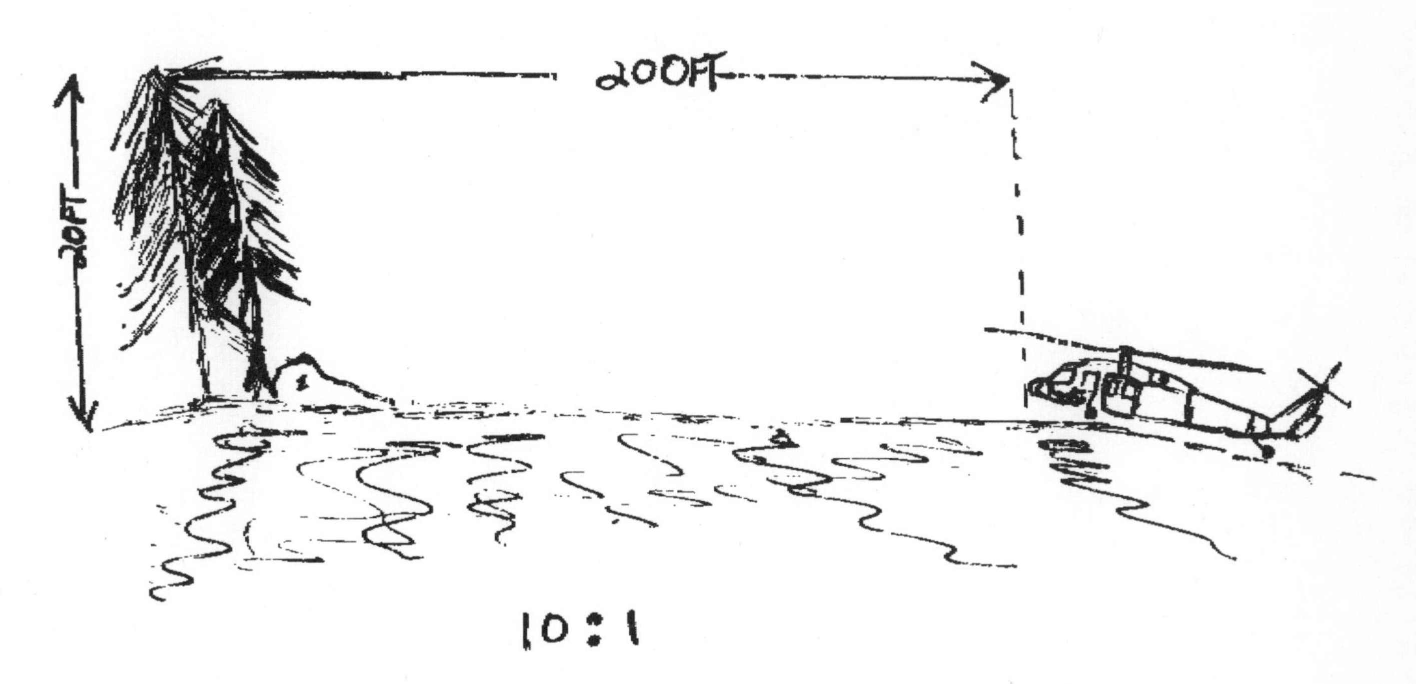
200FT
20FT
10:1
OBSTACLE RATIO

HELO INSERTIONS/EXTRACTIONS
CALL FOR FIRE HELO/SPECTURE C-130

GENERAL. The selection of a landing zone should be selected and prebriefed before the sniper team is inserted into the field. Due to the size of a normal sniper team and the lack of fire power, the sniper team leader should plan for a primary HLZ site and for alternates HLZ sites for extraction along their route of movement incase of overwhelming enemy contact.

1. SELECTION OF A LANDING ZONE.

a. Possible (tentative) locations.

(1) Objective rally points.

(2) Check points.

(3) Pickup points.

b. Factors to consider:

(1) Space requirement.

a. Covered by the supporting units S.O.P.

2. SIZE OF THE HLZ SITE.

a. 25 - 75 meters cleared area.

b. 10 - 1 obstacle clearance ratio. Depending on the density altitude and the load the aircraft is carrying at the time, (a aircraft can insert and extract from a hovering position).

3. NUMBER OF HELOS TO BE USED.

GENERAL. The ideal situation when planning for helo support, is to use insertion/extraction aircraft along with supporting gunships.

a. Cold HLZs. The insertion/extraction aircraft or aircrafts will land at the predesignated touch down points with the supporting gunships entering a standby traffic pattern to the rear of the insertion aircraft so not to enter fear with the departure of the other aircraft, but close enough to engage any unsuspected enemy threats.

B. Hot HLZs. Again the insertion aircraft or aircrafts will land at the predesignated touch down points with the supporting gunships taking station ahead of the other aircrafts. In the situation where there are no supporting ground troops, the gunships will enter a circular traffic pattern engaging any enemy threats to the inserting aircraft and will maintain this pattern until the enemy threat to the aircraft and ground troops has been neutrualized or until the gunships run out of ammunition.

c. Hot HLZ with supporting troops on the ground. The gunships again take station ahead of the extraction aircraft. Prior to the extraction aircraft touching down at the designated point, the supporting gunships and ground troops are used to prep the HLZ by gaining fire superiority over any enemy threats. When the aircraft starts it's flare to land, the ground troops will cease fire and move to the extraction aircraft. The door gunners will take up the ground troops fire, to help suppress any enemy threats to the aircraft and friendly personnel while the gunships maintain station to help neutrualize any enemy threats.

NOTE: If a sniper team is compromised during a operation, one method to break contact would be a helo extraction. Normally a sniper team does not have enough fire power to secure a HLZ site, one method to assist the sniper team in securing the HLZ would be the use of helo gunships as stated above. If there are no gunships to support the sniper team, the team would be forced to move to a predesignated HLZ site along their line of march ahead of the persuing enemy force. The sniper team will have to coordinate their exact arrival to the touch down point and the landing of the extraction aircraft to prevent it from being shot down.

4. SURFACE CONDITIONS. HLZ & PZ.

a. Loose debris.

b. The ground should be firm enough to support the weight of the aircraft.

5. GROUND SLOPE.

a. Utility and observation aircraft will not land if slope exceeds 7 degrees.

b. Large utility and cargo type aircraft will not land if slope exceeds 15 degrees.

c. Never land a aircraft down slope. The tail rotor will dig in when the aircraft flares to land.

6. APPROACH AND DEPARTURE.

a. Aircraft should be landed into the wind whenever possible.

b. Crosswind max. 10-20 MPH depending on the size of the aircraft.

c. Tailwind max 5-10 mph depending on the size of the aircraft.

7. Marking HLZ

General. Marking of a HLZ site will depend on the following:
- No. of Aircraft
- Terrain
- Enemy threats
- Weather
- Unit S.O.Ps

a. There are several methods of marking an HLZ touchdown point
- Land features
- NATO Y (night)
- Single I.R. strobe light (night)
- Smoke (day)/AC panels
- Pyro flares

1) The advantages of using the NATO Y, is that it offers more control of the extraction aircraft from the troop's on the ground.

2) In rough terrain this is the most preferred method of marking a HLZ touchdown point.

3) The disadvantages of using the NATO Y is it takes more time to set up than using a I.R. strobe light.

B. Single I.R. Strobe Light. The use of a single I.R. strobe will again depend on the following:

- Terrain
- Enemy threats
- Weather
- Unit S.O.P.

1) The advantage of using a strobe light is that it takes no time to set up.

2) This method is the preferred method to mark a Hot HLZ touch down point.

3) Troops on the ground who are taking fire can mark their position with a strobe light. Once the extraction aircraft has identified their position, the troops on the ground can guide the extraction aircraft to what ever location they have selected for their HLZ touch down point. This can be done by landing the extraction aircraft on the strobe light or off setting the touchdown point from the strobe light.

4) The disadvantages of using a single I.R. strobe light is it offers NO control of the touch down point. In rough terrain this could present problems to the aircraft.

C. Smoke. Smoke is normally used to mark the touch down point or the smoke can be used to off set the touch down point.

D. Other methods of marking a touch down point are:

- 40 MM Illumination Rounds
- Pop Flares
- Fires
- Land Features
- Imagination

1) For a pilot to identify a HLZ touchdown point, he must be given a recognizable known point to work from, after which a bearing, distance, and discription of the touch down point must be given.

NOTE: IF AIRCRAFT CANNOT I.D. YOUR POSITION SEND UP ANOTHER SIGNAL HAVE THE AIRCRAFT I.D., GIVE THE AIRCRAFT ANOTHER CLOCK DIRECTION AND DISTANCE FROM HIS PRESENT POSITION TO YOURS.

IF GUNSHIP MISSES HIS TARGET OR ENGAGES THE WRONG TARGET TELL HIM TO SHIFT FIRE, THEN GIVE HIM A CARDINAL DIRECTION OF MAG. BEARING AND DISTANCE FROM HIS LAST IMPACT OF HIS ROUNDS FIRED TO THE CORRECT TARGET.

WHEN USING I.R. STROBE LIGHTS TO MARK YOUR POSITION AT NIGHT WHEN THERE IS GROUND FIRE IN THE AREA, THE AIRCRAFT PILOTS CANNOT DISTINGUISH A I.R. STROBE LIGHT FROM GROUND FIRE.

2. EXAMPLE FORMAT FOR CALL FOR FIRE FOR SPECTURE C-130 GUNSHIP.

GUNSHIP A C
GROUND TROOPS G T

AC > GT

OVER

GT > AC

OVER

ROGER AC FIRE MISSION

OVER

GT STATE FIRE MISSION

ROGER AC

ENEMY A.P.C. IN TREE LINE, FROM T.R.P. AW001 BEARING 360 MAG, 700

METERS, THERE ARE NO GROUND THREATS, REQUEST 105 WITH 40 MM.

CAN YOU I.D.

OVER

ROGER GT.

NOTE.

WHEN USING SPECTURE C-130 GUNSHIPS, DUE TO THEIR CONSTANT ORBIT AROUND A TARGET AREA.

OVER

34 > 47

OVER

ROGER 47

WE ARE AT YOUR 3 OCLOCK, 1200 METERS, SIGNAL OUT, CAN YOU I.D.

ROGER 34 I.D. I.R. STROBE LIGHT

ROGER 47

3. ZONE BRIEF ELEMENTS:

a. I.D. CALL SIGN

b. ACTION REQUEST (MEDEVAC, EXTRACTION).

c. GROUND LOCATION (GRID, CHECK POINT, O.R.P., HLZ PREPLAND).

d. WIND DIRECTION AND SPEED.

e. DESCRIPTION OF HLZ AND TOUCH DOWN POINT (SIZE, SECURE/UNSECURE, HOT/COLD, SURFACE CONDITION, SLOPE).

f. DIRECTION OF APPROACH AND DEPARTURE (CARDINAL OR MAG. BEARING).

g. FRIENDLY POSITIONS (IN RELATION TO HLZ AIRCRAFT LAND HEADING USE CLOCK DIRECTION).

h. OBSTACLES IN APPROACH AND DEPARTURE PATH, DESCRIPTION, AND HOW MARKED).

i. TIME AND DIRECTION OF LAST ENEMY FIRE. (CARDINAL DIRECTION AND DISTANCE).

j. SUSPECTED ENEMY POSITION.

k. DIRECTION ENEMY FIRE MOST LIKELY.

l. DIRECTION AIRCRAFT IS CLEARED TO FIRE.

m. LANDING ZONE MARKINGS.

NOTE: SPEAK QUICKLY, CLEARLY AND ACCURATELY TO THE AIRCRAFT, SEND THE ENTIRE BRIEF WITHOUT INTERRUPTION.

7. CALLING FOR AIRCRAFT/COMMUNICATIONS

A. AIRCRAFT CALL SIGN - VH47 (HELO GUNSHIP)
PERSONNEL REQUESTING AIRCRAFT - AQ34

b. EXAMPLE FORMAT CALL FOR FIRE HELO GUNSHIP.

WH-47 > AQ-34

OVER

AQ-34 > WH-47

OVER

ROGER 47 - FIRE MISSION

OVER

STATE FIRE MISSION 34

OVER

FROM MY POSITION NORTHEAST, 600 METERS TANK IN OPEN WITH GROUND TROOPS.

MAKE YOUR GUN RUN, SOUTHWEST, TO NORTHEAST. DIRECTION OFF PULLOUT WILL BE TO THE EAST

I WILL MARK MY POSITION, YOU IDENTIFY

OVER

ROGER 34, I COPY, FROM YOUR POSITION, NORTHEAST, 600 METERS, TANK IN OPEN WITH GROUND TROOPS, DIRECTION OF GUNRUN SOUTHWEST TO NORTHEAST, DIRECTION OF PULLOUT TO THE EAST,

2 MINUTES INBOUND YOUR POSITION

(YOU SPOT AIRCRAFT)

47 THIS IS 34 WE ARE AT YOUR 9 O'CLOCK, 1000 METERS, SIGNAL OUT CAN YOU IDENTIFY?

ROGER 34,

IDENTIFY 40MM PARA FLARE ON GROUND,

ROGER 47, CAN YOU IDENTIFY TARGET,

ROGER 34

CLOSE AIR SUPPORT
(FAST MOVERS)

1. PLANNING AND EXECUTION: A successful air strike begins with a well considered, simple, coordinated plan.

2. TARGET SELECTION:

GENERAL: Effective air support requires a very high degree of accuracy. The distructive radius of conventional weapons is relatively small, so the most suitable air support targets are correspondingly small. Area targets can be attacked by fighters, but this is seldom an efficient use of tac air. More effective results can be obtained by identifying the attacking critical elements within the target area. Proper timing, though difficult to achieve, is a vital importance with many targets. A sudden air attack against an enemy element engaged in assembly, assault, or withdraw can reduce that element to an ineffective fighting force.

3. SELECTION FACTORS.

a. CAPABILITIES OF ORGANIC WEAPONS. Are organic and supporting weapons unable to produce desired results? This does not mean that organic weapons must always be used first. If the job is obviously one for tac air, then ask for tac air. If the job can be done as well by organic means, use them.

b. IDENTIFICATION OF TARGETS. Can the pilot identify the target? Can you pinpoint if for him.

c. AIRCRAFT ARMAMENT CAPABILITIES. Can the aircraft armament achieve the desired results?

d. TIME AVAILABLE. Will the target remain a target long enough to place a strike on it? Can the ground operation afford to wait if there will be a delay in obtaining a strike?

e. CAPABILITY TO DIRECT AN AIRSTRIKE. Can the airstrike be controlled or directed on target?

f. GAIN TO BE REALIZED. from attacking the target as compared with the expenditure of air assets required to do the job.

4. TARGET CATEGORIES. Most close air support targets can be placed in one of the following categories according to their vulnerability to conventional weapons carried by tactical fighters.

a. VEHICLE AND EQUIPMENT. Small items of equipment are highly vulnerable to air attack by most conventional weapons. Wheeled vehicles are easy prey for attack by straffing, rockets, cluster bombs and GP bombs.

b. TROOPS. Troops in the open are highly vulnerable to air attack with all types of ordnance. Protected personnel are much less vulnerable, regardless of ordnance. A man in a foxhole may survive a 500 pound bomb 10 meters away.

c. FORTIFIED URBAN AREAS. Urban areas, as such, are not good tactical fighter targets, specific installations or fortifications within an urban area can be readily destroyed or neutralized by precise attacks with bombs, and particularly with guided weapons.

d. FORTIFICATIONS. Heavy fortifications of bunker or pillbox type are vulnerable only to very accurately placed GP bombs and guided weapons. FAE bombs may be more effective in some cases. Field fortifications without overhead cover such as foxholes and artillery or mortar emplacements are vulnerable to CBU, and accurately placed bombs. To lesser extent, rockets or gunfire can be effective when delivered from higher angles.

5. ORBIT POINT FUNCTION. An orbit point is simply a point in space, usually 10 to 30KM behind the FEBA, established to facilitation orderly flow of air-craft to various parts of the battlefield. Attack aircraft establish radio communications with the FAC as they approach the orbit point (usually one to five minutes prior.) They remain in the vicinity until the FAC completes his attack briefing and clears them to depart.

6. SELECTION. The orbit point may be recommended by the FAC, but responsibility for its selection remains with the strike flight commander. The FAC can exclude certain areas from use due to tactical considerations. The location must permit positive radio communications.

a. The orbit point location is affected by such factors as enemy air defense positions, other tactical operations, and weather. If surprise is a major consideration, the orbit point should be out of sight and hearing of the enemy. The enemy radar warning capability should be considered.

b. If possible, the orbit point should be a position from where the attack can be started.

c. The orbit point must not be in the path of artillery fires or the pilot must be instructed to stay above the maximum ordinate of these fires.

7. HOW TO PINPOINT THE TARGET.

GENERAL. A rule to remember is that if a pilot can see the target, he can usually hit it. The target must be identified to attack pilots as clearly and precisely as possible. The supported ground unit must communicate the target location to the FAC and, through him, to the attack pilots. An airborne FAC may "talk" the fighters to the target, or mark the target with

smoke rocket, or he may request a mark from the ground. Without an airborne FAC, a mark from the ground will usually be necessary. In every case, a clear visual mark will reduce the chance of confusion or misunderstanding.

METHODS. Each of the many ways of pinpointing a target have advantages and disadvantages. A combination of methods is often best.

a. TARGET COORDINATES. Their value in actually pinpointing the target for strike will vary. Usually grid coordinates alone are adequate only for area type targets. Accurate 8 digit coordinates are necessary for radar bombing. Six digit coordinates are acceptable for visual strikes, the jet pilot usually be unable to pinpoint a 6 digit UTM coordinates due to his speed, altitude, and map scale (1:250,000 or 1:500,000). On the other hand some of the same aircraft have inertial navigation systems that automatically provide the pilot with latitude and longitude coordinates to the nearest second (100 feet). LORAN systems also provide 8 digit UTM coordinates. Errors build up in inertial systems with time if they are not coupled with the LORAN systems for continuous updates. If a pilot is provided with the latitude and longitude or a 8 digit UTM coordinates of a nearby prominent terrain feature, he can update his inertial system and obtain accurate information for several minutes.

b. GEOGRAPHICAL LANDMARKS or terrain features, clearly visible from the air, can assist greatly in target identification when used with another location method. Streams, roads, bridges, tree lines, cultivated areas, prominent hills, etc., help narrow the area the pilot has to search, i.e., "in the tree line on the north side of a big square wheat field," means the pilot only has to look in the tree line on the north side of the big square wheat field.

c. REFERENCE POINTS are the most common aid to visual location of a target. The pilot's eyes are led to the reference point and from it to the target, sometimes through a series of decreasingly obvious reference points. This is done usually in the following manner:

1. Cardinal directions (NW, SW)

2. Magnetic compass bearing (degrees not mils)

followed by a distance from the reference point to the target. Reference points (TRPs) similar to those used to lay artillery fire to precisely identify a target can either be preplanned or may be spontaneous references. Methods to mark a target are as follows:

1. Smoke rounds from mortars, artillery, or grenade launchers (40MM), are the primary target marks. White phosphorus (WP) is usually the best because the cloud blossoms quickly and is highly visible. The round can be timed to the impact when fighters are in the best position to attack. Another method that is most commonly used is a 40MM parachute round with the parachute cut away and extracted from the round itself leaving only the pyrotechnic, this is a excellent method to small ground units in the field.

2. Tracer fire can be used to mark a target at night. (The target is located at the intersection of the two streams of tracer fire or the impact point of a single stream. You should be able to order your gunners to commence or cease fire in rapid response to pilot request. Tracer burn is limited.

3. Ordnance already impacting on the target may provide an adequate mark or reference. There is a possibility of confusion if resulting smoke, fire, and etc., look like other nearby smoke or fires that the ground observer is not aware of. Where there is any chance of any confusion, a coordinated marking round should be used.

4. Illumination rounds.

5. Grass fires.

6. Friendly positions, when clearly recognizable from the air, may be used as day or night reference points for the location of close-in targets. If a position is large, directions to the target should be given from a specific visible feature within the position (i.e., I.R. strobe, chem lights, pop flares, smoke, pannels, etc.,) the ground observer should be located near the mark or ground feature, if not the pilot should be informed of his relative position.

7. Laser target designator will allow forward-deployed ground personnel to "mark" targets for tactical aircraft and to designate for the delivery of laser-guided weapons. This system is primarly a good weather, day or night spotter.

8. Radar beacons

9. HOW TO IDENTIFY FRIENDLY POSITIONS.

1. When to mark. Friendly positions should be always marked during close airstrikes if there is no danger of compromise to ground troops. In some situations it may be necessary to accept compromise in order to conduct a safe airstrike. In other situations it may be adequate to say "all friendlies are south of the target" 500 meters. As a general rule a mark is required when troops are located closer then 300 meters to the target. Regardless of the situation the pilot feels better when he knows the location of friendly troops.

10. ATTACK HEADING SELECTION.

GENERAL: Pilots like to use random attack heading whenever possible in order to confuse enemy anti-aircraft gunners. It is also desirable to attack along the long axis of the target for maximum affect. The FAC may however, have to restrict attack headings in the interest of safety when there are nearby friendly positions or when the aircraft turn heading is directly over enemy positions. (The FAC should advise the aircraft of the enemy situation on the ground, where the threat is located, description, direction of aircraft pullout, location of friendlies, attack heading, type ordnance needed, aircraft pull out direction will be given towards friendly position not the enemy to prevent the attacking aircraft from being shot down.)

1. You are safest if the attacks are parallel to your front.

2. Attack toward your position is undesirable because of possible ricochets and ordnance early release.

3. Attack over an attack from behind and over friendly positions are also undesirable because of the dumping of empty cartridges overboard as they strafe creating a missle hazard.

11. <u>REQUESTS AND COMMUNICATIONS.</u> Airstrike requests can be either preplanned or could be requested from ground troops in the field for immediate close air support.

1. <u>ESSENTIAL ELEMENTS OF A REQUEST.</u>

a. Requester's identification. (call sign).

b. Request type and priority. (state whether the request is for a pre-planned or immediate mission, also designated the priority of the request. (priority is only essential in cases where several requests have been submitted.)

c. Target description.

d. Time on target. (Indicate the time desired and the latest time acceptable if it is not immediate request.)

e. Desired ordnance.

f. Target location.

g. Direction of pullout.

h. Friendly's position.

i. Aircraft attack heading.

j. Situation on the ground, location of enemy, friendlies.

C-2 CALL FOR FIRE

a. Normally, the call for fire will go through the forward air controller (FAC). In the event the FAC is absent, ground personnel may direct strike flights onto targets. Corrections to the target must be simple, clearly understood, and fast (timely). Cardinal directions are preferred over clock reference or attack heading corrections.

b. The observer-target method of correcting mortar or artillery fires could be dangerously confusing in a fast moving air strike. For example, a forward air controller would tell a pilot to place the next burst 300 meters NORTH of the previous rounds rather the RIGHT 300.

(1) Observer identification and location. Use smoke grenades. Avoid use of red or white smoke to indicate friendly positions as these are used to indicate enemy positions and may draw fire.

(a) Notify the pilot that you have thrown smoke, let him identify the color, then confirm the identification.

(b) Colored panels may be used to identify friendly troops and give the pilot the general direction of attack. Ensure identification.

(c) Any expedient means such as T-shirts, tracer rounds, or flaming arrows may be used.

(d) Remember, any method you use to mark your location can also reveal your location to the enemy. Use caution! Try to show yourself only to the friendly aircraft whenever possible.

(2) Methods of target identification. Ensure that the pathfinder and pilot have the same map and coordinate system. Targets may be identified by:

(a) Grid coordinates.

(b) Reference to prominent terrain features (natural or man-made).

(c) Colored smoke fired from previous aircraft or from mortar, artillery, or recoilless rifles.

(d) Simulated attack runs may be made if the pilot is not sure of the target. He attacks the position he thinks is the target but does not expend ordnance. From this you can adjust by verbal command to move him on target. (Avoid this method when possible as it exposes the aircraft to the enemy. Use only when absolutely necessary.)

(e) Reference to observer's location (i.e., polar coordinates consisting of a cardinal direction and distance).

(3) Pertinent information about the target.

(a) Target and friendly troop separation distance.

NAVAL SPECIAL WARFARE

SCOUT/SNIPER SCHOOL

DATE________________

EVASION

INTRODUCTION

1. This is designed as an aide to Evasion. It does not constitute rules. The necessity to evade in the present day could occur from:

a. A breakout from PW Camp.

b. A breakout from a surrounded position in small numbers or as an individual.

c. As a result of tactical nuclear strikes and the eventual disorganization of Corps and divisional boundaries.

MOVE BY DAY

2. Moves whether of individuals or groups must be planned in advance. Moving by day is inadvisable, but sometimes unavoidable i.e. after a PW breakout or when an extremely long distance must be covered.

MOVE BY NIGHT

3. 90% of evasive moves should be by night. But darkness often breeds over confidence. There will be a compromise between taking the easiest route and avoiding going where the enemy expect you. Study and memorise your route in order to avoid using light to map read.

4. Never move on roads. If crossing a road, locate sentries and if necessary use a diversion. (Cross immediately after a vehicle has passed, noise and light.)

5. Never cross bridges. Try improvising rafts in order to keep clothing dry, or swim.

6. In hills avoid using ridges as you are likely to be silhouetted and remember you can be seen from below for a greater distance than you can see. After crossing a skyline change direction on a downwards slope and look behind to see you are not being followed.

7. Keep away from population of any kind. ALWAYS have at least one emergency RV. Know how long it will be open. When you are making for RV after enemy contact, make sure you are not followed.

8. Use a leading scout as far forward as possible even when only two men are together.

9. Avoid walking in mud, through standing crops or any place where obvious tracks will be left. Leaving litter or any signs of occupation in a lying-up area is asking for trouble.

10. Danger Zones. The following points will help evasion in dangerous areas:

a. Cordons. These are relatively easy to pass at night. If you watch for up to 2 hours some enemy soldier will give away his position by noise, movement or normal sentry relief. Once a position is located pass as near to it as you safely can.

b. Cordons will nearly always be near roads because enemy transport can be quickly deployed off them. This will not however be the case if the enemy have available helicopters in quantity. If they are heard expect cordons to be in low ground or to use flares from high ground. Para 6 above becomes very important.

c. Cross roads immediately after vehicle using light which has passed. These will blind enemy sentries who seldom, if ever shut their eyes to the light.

d. Imitate silhouette of enemy sentries for so far so possible. In particular headgear. Learn at least one phrase in his language, but you must be able to say it fluently.

LYING UP POSITION (LUP)

11. Selection. Do not use isolated cover, particularly if it is marked on a map. A thick hedge or long grass if often better than small woods.

12. Entry

a. Whenever possible after dark.

b. Be careful not to leave tracks If possible re-organize position at first light.

13. Siting

a. Concealed from ground and air.

b. If possible only one good approach.

c. Easy escape route.

d. Near water if you intend to stay more than one day. Otherwise take water in with you during the night.

e. A good location for an LUP would be long grass, vegetation or scrub in an isolated position.

14. Procedure in LUP

a. Keep quiet and still.

b. Have a sentry if in a group of more than two.

c. Bury all refuse.

d. Kit always packed and if in possession of weapons, clean one at a time.

e. Men always ready to move quickly i.e. compass, rations, map on body. Weapon at hand.

f. Emergency RV must be known and withdrawal route planned.

g. Before evacuating site search for any avoidable trace of occupation.

h. Smoking must be controlled i.e. smoke showing by day, cigarette end glowing by night.

PARTISANS OR AGENTS

15. There are basically two types of contacts an escapee can make:

a. An organized contact after a PW breakout, with prior knowledge of RVs.

b. A chance contact, not previously planned, with a reliable source i.e. a doctor or priest in an enemy occupied area.

16. The civilian agent if caught has more to lose than you so after making contact:

a. Make up your mind to trust or distrust him.

b. Ensure RVs are secure and that you have a drill at them i.e. one man entering before remainder when in a group.

c. Do all the agent says, but never say who previous contact was.

d. In the case of 16(b) ensure that he is alone before contacting.

e. Have an emergency RV in case something goes wrong.

17. It is the personal determination of the escapee which will ensure his success. Compliance with the above principles will only serve to make the task easier.

NAVAL SPECIAL WARFARE

SCOUT/SNIPER SCHOOL

DATE ____________________

DOG EVASION

INTRODUCTION

Man has used the dog for military purposes for thousands of years. The Egyptians, Huns, Romans, all resorted to the use of Guard and Tracker dogs and no doubt the evasion tactics employed then have changed very little. Henry VIII provided Spain with large attack dogs, wearing spiked collars, to fight the French.

The availability of chemical aids is limited. With the current trend of interest shown by many Governments, some progress will be made in this field, but as in all research, finance, and more pressing needs must take precedence.

It is also possible to produce chemical aids for the handler and his dog to overcome evasive aids. The result could be the evasive aid becoming a beacon for the dog to home on.

These very general notes are therefore written for the guidance of personnel who find it necessary to evade working dogs and in so doing have no chemical aids available.

If you are supplied with such chemical or mechanical aids, use them as an addition to your evasion technique and not as a replacement.

The dog used for Military purposes must conform to certain requirements, irrespective of its breed. These can be summed up as follows: -

Physical	Height in shoulder 22 to 26 inches Weight varying from 45 lbs to 100 lbs plus Speed in excess of 25 miles per hour
Temperament	Intelligent, Courageous, Faithful, Adaptable, Energetic.

There are many breeds having these requirements, such as Alsation, Dobermann, Pinscher, Rottweiler, Mastiff, Boxer, Collie, Groenendael, Schnauzer (Giant), Labrador etc.

The breed of dog employed at a particular base may be varied to suit the climatic conditions under which it will work. Humidity and temperature being the main factors involved.

It's attention is, however, drawn by movement and if it's interest is roused, will follow up with hearing and nose.

Dogs have nonochrome vision, with a limited depth of field. There appear to be areas at certain distances where focus varies. As in humans, vision varies from dog to dog, as does the inclination to use sight.

At night the dog is able to detect movement, due mainly to its low position looking up at the skyline. It makes more use of what light is available.

SOUND

With a range of hearing twice that of humans, the dog is attracted by noise not received by the handler. Beware of equipment rubbing together, radio equipment, burners, etc. The distance at which received is very much affected by weather in particular wind and rain. Obey the rules of approach from down wind.

Dogs used for military purposes are divided into two basic groups. Those which rely on scent carried in the air and those who rely on scent held on the ground.

The very basic division is applicable mainly to training and there is no doubt that an experienced dog in either field will naturally progress from one scent source to another when the need and the interest is great enough.

However, the division into these two groups is sufficient for evasion purposes. Many rules apply to both. Bear in mind that there can be great variations in requirement from types of dogs for instance - using air scent e.g. Guard, Defense and Search. The same will apply to those using ground scent.

SCENT

The dog's sense of smell is many thousands greater than our own. Through it's olfactory organs it has the ability to detect a source of scent, either by following air currents, or tracks left on the ground. This natural ability to hunt has been controlled by man, and the search and tracker dogs have emerged. These dogs must have the physical capability of following such tracks for many miles.

Human scent from a dog's point of view is a combination of smells from many sources.

BODY SCENT

The smell of the human body, made up of 'body odour' produced in abundance by the sweat glands, in particular under the arms, legs, etc. This particular odour is increased by rapid movement, nervousness tension, various types of food and uncleanliness.

To this odour, we must add the following:

Clothing, deodorants, toiletry, shoe leather, polish, chemical aids if used on clothing, environment (Patrol, oil, timber, etc.), and many other that the human may have been in contact with.

Race and creed play a part in the individual definition of a particular scent.

The amount of total body scent, produced is greatly affected by constitution, activity and mental state.

It therefore follows, that in many respects you can control your flow of body scent. Keep cool, calm and more with confidence.

GROUND SCENT

Body scent deposited by the soles of the feet, plus body scent drifting down, but mainly ground disturbance caused by the weight of the man on the ground.

This contact of the foot produces scent from the following sources. Crushed vegetation, insects, deposits from shoes. The breaking of the surface allowing gas and moisture to escape. All these scents added together produce the main scent for the tracking dog.

Airborn scent is soon dispersed leaving the dog with the ground scent only. An experienced tracking dog can follow this scent up to forty eight hours afterwards, in virgin, humid territory.

The trained tracker dog can find the direction of the track. This is possible because of the purchase of the foot. The toe part of the impression is deeper, and remains in contact longer. After examination of several foot contacts the dog can follow the track in the correct direction.

Because of the natural evaporation taking place on the surface, with variation in moisture and gas movement, the basic content of each track varies from minute to minute. This variation together with the deposited body scent makes every track different. It is this variation and the ability of the dog to compute through its olfactory system the basis of each track, that the dog can follow an individual scent, even when many other tracks are present.

THE TRACKING DOG

The following will give an evader sufficient detail to make a good attempt at tracker dog evasion. As no two dogs react in the same way to a given set of circumstances, we can only generalize. It is for this reason that the notes are in three groups.

1. Before contact with enemy

2. Contact from a distance

3. Close contact

These headings are for convenience only, and any of the acts given can be applied to each quite successfully.

1. Before any contact is made with the Enemy

a. Associate oneself as much as possible with the surroundings. The rules of physical camouflage should also apply to personal scent. Keep in with the surroundings. Alien scents attract the dog.

b. Travel over ground already used by humans or animals.

c. When travelling in groups split up every now and then. This need be for only a short distance, but will be sufficient to slow the dog down.

d. When preparing food, take care as to direction of smoke and fumes. Handle wrapper and containers as little as possible. When burying, do not handle the ground, use metal instrument. If possible, sink in deep water.

e. When entering or leaving L.U.P's, do so from different directions. Make false trails round perimeter of L.U.P.

f. Follow to the side of animal tracks, thereby leaving no footprints.

2. Contact from a distance Visual contact or dog locating track

a. Speed and distance. Tire the dog, destroy handler's confidence.

b. If in group, arrange R.V. Split up.

c. Vary surface and terrain. Where possible use metalled surfaces, cross and re-cross at intervals.

d. Pass through fields which contain, or have contained, animals.

e. When travelling through woods, scrub or brush, change direction frequently. Remember dog will usually be on a line. This becomes easily tangled, and will slow or stop dog for a time.

f. If possible cross streams etc. Walk along streams for short distance and make false exit and entry points. Walking too far in water will slow own progress to much.

g. Take any step which slow dog without further endangering self e.g. false trails, use of roads, entry into villages.

3. Close Contact - Dog in position to be released and able to attack.

a. Get out of sight of handler.

b. Change direction

c. Use metalled, stone, rough surfaces

d. Pass through animals

e. Clear obstacles

f. Shed articles of clothing food etc., any scientific aids

g. Wherever possible try to part handler from dog.

h. If dog catches up with patrol - silent destruction, using tactics as for guard dog.

There are many factors which affect scent, and a dogs scenting capabilities. These factors can best be summarized as follows: -

Favourable - Moist ground conditions
Vegetation, grass fern, etc.
Humidity
Forest areas
Light rain, mist, fog
Slow moving quarry
Quarry carrying heavy burden
Nervous quarry - excess perspiration
A number of persons on the move
Light winds
Still, sturgid water, i.e. swamp

Unfavourable - Arrid
No vegetation
Metalled surfaces, sand, stone
Animal scents, tracks
Motor, factory, pollution
dust, etc. irritating to dogs nose
Quarry continually taking evasive steps
resulting in handler losing confidence in dog.
Ploughed ground
Gale Force winds
Ice, snow, water

THE GUARD DOG

The larger breeds of dog are used for this purpose. The final objective being to chase and attack. It must have the courage, and physical capability to fulfil the objective. It is useful to note that various methods of training are employed throughout the world, varying from compulsion to revulsion. Irrespective of training design, the end product is basic - Attack and Detain.

The guard dog is operated in two ways; with a handler on leash or roaming free in a compound. Whichever method is employed, the dog will rely primarily on its hearing and scenting ability to detect intruders. It's sight, being less developed, will be used as an auxilliary detection, the dog being drawn to a particular area by movement.

After detecting an intruder, the dog will operate on command of the handler or on situation stimulas. The handler command is normal, but the situation stimulas is where a dog is released into a compound and will attack any person entering, other than a known guard or collection vehicle. Some dogs are so trained that any person is attacked, it being necessary to collect

directly into a cage within the compound. Here the basic command to attack is the physical presence of a human being.

In either case the dog will retain its grip on its quarry until ordered to leave. In the case of highly aggressive dogs, strict compulsion may be necessary.

It is this courage and ability of the dog that makes it vulnerable to the intruder. Pad oneself as described below, encourage the dog to attack, biting in a place that you dictate. Present a target to the dog, thereby placing it in a position in which it can be immobilized or destroyed.

Adequate protection can be had from wrapping round the arm any of the following, webbing belt, leggings, rifle sling, ponchos, wrapping from equipment, scarves, headgear. Always have a layer of softer material inside and outside your main protection. The inner layer to take some of the pressure, the outer to give the dog something to grip on.

The dog is far less dangerous if it makes firm contact on the first run in. If it falls off or is deterred, it will look for an alternative target and then begin to dictate the situation to you.

Throughout its training the dog has always been allowed to succeed. It is this inbuilt confidnece in its own ability that encourages the dog to overcome every obstacle. Give it the opportunity to succeed and then destroy. It is most vulnerable when gripping target.

Remember a dog deterred will bark or growl, drawing the attention of the guards.

To avoid initial detection, obey the following simple rules: -

1. Always approach from down wind.

2. As silently as possible.

3. Ensure you cover the last part of the journey as slowly as possible thereby cutting down excretion of body odour.

4. Keep all garments securely fastened. Where a draw cord is fitted, keep it tied.

5. If you have to stop for any reason before entering the perimeter do so outside the 200 meter mark. Within this distance dogs have detected intruders travelling against the wind as well as with the air flow.

6. Keep as low as possible, use natural hollows. The air scent will be obstructed by undergrowth or barriers.

7. Be aware of changes in scent direction caused by barriers, i.e. around buildings.

8. Approach from an area where you know other humans operate in, or approach from. The dog pays less attention to areas where it expects there

to be persons or vehicles. It may be attracted, but under some circumstances, this identification will be misinterpreted by the handler.

9. When within the perimeter fence, remember, the dog relies mainly on sound and scent. Its attention will be drawn by movement. If you are down wind and the dog is passing, keep still. Guards have passed within 10 yards without being attracted.

10. The average guard dog will have difficulty in detecting persons up high. If they do, they have difficulty in pinpointing locations. This delay will give you time to operate.

DESTRUCTION

The destruction of a trained dog is by no means a simple matter. The situation is made more difficult for the evader, by the necessity for silence, or at least a degree of quiet.

It is often easier to take the dog and immobilize, by either tying to a secure fitting, or binding the front legs. Always muzzle, and if possible render it inopperable, example, breaking a leg.

Actual destruction may be by any of the following: -

1. Stab through abdomen, aiming from rear to front.

2. Sticking pointed stick, spear into abdomen

3. Severe blow to skull

4. Shooting through skull, aiming above, and in centre of line drawn diagonally from ear to eye.

5. Shooting through back

6. Chop at back of neck just before shoulders

Whichever method is decided upon, supreme physical effort must be exerted. The dogs skeletal system is such that is is virtually armour plated. Go for the soft spots, the abdomen, or the point beneath the chin, and above the brest bone.

THE SEARCH DOG

This dog, trained to quarter an area, with minimum command. On location of an intruder, to give tongue, or return and collect handler and patrol.

Relies mainly on locating source of air borne scent. Make sure that you keep that source as small as possible.

When in an L.U.P. observe the following: -

1. Keep as close to the ground as possible.

2. Have the majority of clothing over you, let the earth absorb the scent.

3. Breath down into the ground, or at least into low vegetation.

4. Keep still.

5. If burying items, do so underneath your lying point, all smells kept down by body and covering.

6. Restrict smoking, fires, etc. Dogs whilst searching are drawn by any alien scent.

7. This type of dog is more inclined to circle and bark, or collect handler. Depart when possible and use normal evasion techniques.

8. In all circumstances if located, and escape not possible, catch and destroy.

Remember always, that the dog be it guard, search or tracking, is reliant on command from a handler. These commands may be by voice, whistle or hand signal. They may not be continuous, or obvious, but are always necessary. It is this reliance of the dog on the human that makes an opening for the evader. Part them, and the dog begins to lose confidence. Change the dog's surrounding and immediately its sense of security is weakened.

Always aim to: -

a. Destroy the confidence of the handler in his dog.

b. The confidence of the dog in the handler.

c. Confidence in themselves.

CONCLUSION

There are many and varied opinions regarding evasion. This state of uncertainty is due mainly to the very limited amount of proven information we have of the dogs interpretation of scent, and it's ability to distinguish between scents.

As humans we tend to base all theories on our own standards, thus expecting the dog to live up to our requirements.

The dog does not have the capability to penetrate the human mind, although there may be a transference of feeling. We, on the other hand, can study the psychological qualities of the dog, and understand him. In so doing, discover his weaknesses, and his vulnerable points.

AN/PRC-117

Lesson purpose: To familirize SEAL scout sniper in the operation of the AN/PRC-117.

Objective: To familirize the SEAL scout sniper in the capiblities and operation of the AN/PRC-117 for future field use.

LESSEN OBJECTIVE:

a. To familirize the SEAL sniper in:

1. Equipment setup.

2. Equipment characteristics and capabilities and features.

3. Radio set characteristics and capabilities.

4. Description and use of operators controls and indicators.

TRAINING OBJECTIVE:

To Train the SEAL scout sniper in the required skills needed to operate the AN/PRC-117 in a field enviroment.

AN/PRC-117

GENERAL:

Frequency   30oo to 89.975 MHz.

Channel spacing 25 KHz.

Preset channels   8 programmable.

Number of channels   2400.

Modes of operation:

1. Narrow band voice.
2. Wide band , data (to 16K bits/sec).
3. Retransmit.
4. Simplex or half-duplex.

Battery   BA-5590 or   nickel cadmium, rechargeable.

Battery life 20 hours at 1 watt, 12 hours at 10 watts.

Weight   12.75 lbs. (includes battery, anttenna, handset).

1. <u>OPERATION OF THE RADIO SET.</u>

a. <u>Turn all controls up</u>.

This easy to remember rule gives the simplest starting postion when picking up the radio.

b. <u>Push TEST/LOAD.</u>

This initiates self-test. The L.E.D. display shoes the status of the radio set:

(1) NORMAL - Battery voltage displays, then goes blank.

(2) FAULT - Battery voltage displays, then "A" and number of faulted module.

NOTE: Use self-test as often as needed during radio operation to check battery voltage and overall radio function. Self-test can be used in any mode except PRGRM and RMT.

c. Set controls as follows:

(1) Channel - Select MANUAL or PROGRAMMING FREQUENCY - 1 - 7.

(2) Mode - Select a SQUELCH mode for transmission/reception:

* OFF - No squelch.
* NOISE - Receiver squelched untill carrier is detected.
* TONE - Receiver squelched until carrier is detected.

NOTE: These squelch modes are fully AN/PRC - 77 compatible. A radio set for TONE SQUELCH mode can only receive from a radio transmiting the 150 Hz subcarrier tone.

d. Push to talk.

2. SETTING FREQUENCY-MANUAL CHANNEL. (SIMPLEX MODE)

NOTE: Simplex mode is when the radio set is transmitting and receiving on the same frequency.

NOTE: Half-duplex mode is when the radio set is transmitting and receiving on différent frequencies.

a. Set these controls: (SIMPLEX OPERATION)

* VOLUME - on.
* Channel - MANUAL.
* Mode - any postion other than SCAN or RMT.

b. Push DISPLAY.

c. Select frequency.

* Toggle the MHz and KHz switches UP or DOWN.
* Display blanks automatically after a few seconds.

* To recheck the frequency, push DISPLAY.

* SETTING FREQUENCY - MANUAL CHANNEL. (HALF-DUPLEX)

NOTE: Do sets d-g only if transmit frequency differs from receive frequency.

NOTE: For half-duplex operation on the manual channel, first enter the receive frequncy using steps a-c. Then install the handset and do these steps:

d. Key the handset (KEEP KEYED UNTILL STEP g).

e. Push DISPLAY.

f. Select new transmit frequency displays.

* Toggle the MHz and KHz switches UP or DOWN.

g. Release the handset key.

* Display blanks automatically after a few seconds.

* To recheck the receive frequency, push DISPLAY.

* To recheck the transmit frequency, key the handset, then push DISPLAY.

* PROGRAMING FREQUENCY - CHANNELS 1 - 7. (SIMPLEX MODE)

a. Set these controls:

* VOLUME - on.

* CANNEL - 1 - 7.

* MODE - PRGRM.

Set DISPLAY DIM AND XMT POWER as desired.

b. Push DISPLAY.

* Current frequency displays. Display must be lit for frequency to be changed.

c. Select frequency.

* Toggle the MHz and KHz switches UP or DOWN.

d. With frequency displayed, push TEST/LOAD.

* The display blanks after a few seconds.

* To recheck the frequency, push DISPLAY.

(EXCEPTION: 60.000 display may mean a jumper option prevent recall of frequency. see manul for more information).

* PROGRAMMING FREQUENCY - CHANNELS 1 - 7. (HALF-DUPLEX)

NOTE: For half-duplex opreation on programmed channels, first enter the receive frequency using steps a - d then, install the handset and do steps e - i:

e. Key the handset (keep keyed untill step 9).

f. Push DISPLAY.

* Current transmit frequency displays.

g. Select the transmit frequency.

* Toggle the MHz and KHz switches UP or DOWN.

h. With frequency displayed, push TEST/LOAD.

i. Release the handset key.

* The display blanks after a few seconds.

NOTE: Whenever the receiver frequency is reprogrammed, the transmit frequency automatically reverts to the new receive frequency.

3. ADJUSTING CONTROLS.

a. XMT POWER CONTROL.

(1) Controls transmitting power.

* LOW = 1 WATT.

* HIGH = 10 WATT.

For longest battery lift, set at HIGH only when necessary.

(2) VOLUME CONTROLS.

* Controls on/off and audio volume.

(3) DISPLAY DIM/WSPR CONTROLS.

a. DIM.

* Controls LED display brightness. Turn conterclockwise to dim display.

b. WSPR (WHISPER).

* In transmit, WSPR postion increases audio gain at the mouthpiece 10 dB.(When whispering, full audio power is transmitted.)

(4) FREQUENCY CONTROLS.

* Used to select frequencies. Active only when display is lit.

4. <u>SCAN.</u>

* Using SCAN capability to continuously scan radio activity on all 8 channels.

(1) For SCAN operation, set up these controls:

a. VOLUME - on.

b. Channel - any channel (keying is enabled on selected channel).

c. Mode - SCAN.

d. Set DISPLAY DIM and XMT POWER as desired.

NOTE: ECCM channels 5,6,and 7 are programmed with codes - not discrete frequencies as for channels 1 - 4.

b. Enter the first code as in PROGRAMMING FREQUENCY steps a - d.

NOTE: The upper 2 (MHz) digits of the code determines the 5 MHz ECCM band used by the signal. The right ones (MHz) digit and lower 3 (KHz) digits of the code determine part of the ramdom "HOPPING" pattern for ECCM transmission and reception on this channel

ECCM FREQUENCY BANDS (MHz):

30.000-34.975	60.000-64.975
35.000-39.975	65.000-69.975
40.000-44.975	70.000-74.975
45.000-49.975	75.000-79.975
50.000-54.975	80.000-84.975
55.000-59.975	85.000-89.975

7. ECCM OPERATION.

a. In general, use of ECCM channels no differently from non-ECCM channels. Remenber that ECCm is not operational in SCAN mode.

(1) MODE CONTROL POSTION.

a. Whe radios in a network are set up in the same ECCM codes, they are compatible for ECCM communications in either OFF, TONE, or NOISE SQUELCH modes.

b. Minimum synchronizing information is transmitted when all network radio are set to NOISE or OFF SQUELCH modes.

c. When radios are set to TONE mode for ECCM operation, each keying synchronizes the network.

(2) CLEAR-NET ENTRY.

a. During ECCM operation, use this automatic scanning sequence to respon to non-ECCM calls.

b. When set to an ECCM channel, the radio automatically scans a single clear (non-ECCM) channel.

c. Set to ECCM channel 5, the radio scans clear channel 1. Set to channel 6, it scans clear channel 2. Set to channel 7, it scans clear channel 3.

NOTE: As the radio scans, it blinks the decimal point on the L.E.D. display. When it detects a singal on a channel, the radio displays "C" and the number of the channel, 0 - 7. (exsample: C0 indicates signal detected at the MANUAL channel frequency.) The radio "locks in" a channel at the same time, allowing the operator to hear the recieved signal. The radio continues to lock in a channel as long as the sigal is being recieved. The display blanks after a few seconds.

When contact is broken, the radio contiues to scan, begining with the next channel.

The radio scans at a rate of approximately 10 channels per seconds. When scanning fewer than 8 channels, it may be more advantageous to load the same frequencies more than once - using otherwise unused channels to shorten scan responce time.

5. TRANSMITTING IN SCAN MODE

* The radio allows transmission in the SCAN mode ON CHANNELS SELECTED BY THE CHANNEL CONTROL. The antenna is automatically tuned for transmission on this channel in SCAN mode.

* . When using this feature, remember that the radio will continue scanning in this mode when not transmitting. (The radio could lock in a channel different from the one you are using, tempotarily preventing you from receiving in this mode. Keying the handset puts the radio set back on the channel selected on channel control.)

6. SCANNING ECCM CHANNELS 5 - 7

(1) With the AN/PRC-117 VHF-FM , channels 5 - 7 are dedicated for frequency-hopping ECCM transmission and reception.

NOTE: The radio cannot detect ECCM transmission while in the SCAN mode.

NOTE: The radio does not transmit a freqency-hopping signal when keyed in SCAN mode.

ECCM operations is not compatible with SCAN mode operation because of the unique charcteristics of the frequency-hopping signal.

(2) In scan mode, the radio treats ECCM channels 5 - 7 in the same as other radio shannels. It scans the frequnecy entered in any of these channels for a received signal. It transmits when keyed in channels 5, 6, or 7 but this transmission is not ECCM FREQUENCY-HOPPING transmission.

(3) PROGRAMING ECCM CHANNELS.

a. To program ECCM channels, use steps of PROGRAMMING FREQUENCY.

d. The radio set scans wherever it is actively transmitting or receiving.

e. When a clear channel signal is detected, the radio beeps in the handset and displays "C1", "C2", or "C3" for the detected channel.

f. To transmit or recieve on this clesr channel, set the CHANNEL CONTROL to the CLEAR NUMBER displayed. Otherwise, ECCM operation continues.

g. For the clear channel, use a frequency within the 5 MHz band used by the corresponding ECCM channel. Select TONE squelch on clear channel radios.

AN/PSC - 3

Lesson purpose: To familirize SEAL scout sniper in the operation of the AN/PSC-3.

Objective: To familirize the SEAL scout sniper in the capiblities and operation of the AN/PSC-3 for future field use.

LESSEN OBJECTIVE:

a. To familirize the SEAL sniper in:

1. Equipment setup.

2. Equipment characteristics and capabilities and features.

3. Radio set characteristics and capabilities.

4. Description and use of operators controls and indicators.

TRAINING OBJECTIVE:

TO Train the SEAL scout sniper in the required skills needed to operate the AN/PSC-3 in a field enviroment.

GENERIAL INFORMATION:

The AN/PSC-3 is a portable radio set designed for satelite communications, it also processes the capiblity of (UHF LINE OF SIGHT) communications. The radio set is designed for long range tactical ground to ground and air to ground communications. The radio set can be used with COMSEC TSEC/KY-57 speech security equipment for secure voice communication.

1. Radio set characteristics and capabilities.

(1) Portable.

(2) One band - UHF (line of sight).

(3) Tactical mode.

a. Ground to ground.

b. Ground to air.

(4) Secure voice mode.

a. When used with attaching cable and COMSEC/KY-57.

(5) Preset channels

a. Up to 4 preset channels can be set into radio memory before a mission.

(6) Battery types.

a. 2 each non - rechargeable BA - 5590/U lithium organic batteries.

2. DESCRIPTION OF RADIO SET COMPONENTS.

(1) Receiver - transmitter.

a. Contains controls, indicators and electronics to operate the radio set in SAT-COMM. and UHF line of sight tactical modes.

(2) Battery case.

a. Houses the batteries.

(3) Batteries.

a. Supply power for operation of the radio set.

(4) Handset.

a. Provides audio input/output for the radio set.

(5) UHF antenna.

a. Used during Line of sight operation.

(6) SAT-COMM antenna(DMC-120).

a. Used during satelite operation.

(7) KY-57 baseband cable assembly.

a. Connects between RT. AUDIO connector and KY-57 unit during secure voice operation.

3. OPERATING INSTRUCTIONS.

GENERAL:

Controls, indicators and connectors used by the operator of the radio set are discussed in two groups.

a. Display.

b. Controls and connectors.

1. Display.

(1) The display indicates frequency entered in the radio set.

1. Controls and connectors.

a. Volume switch.

(1). Power on/power off, adjust the volume of the radio.

b. Function switch.

(1). Manualy sets the radio set to SAT-COMM. or LINE OF SIGHT operation.

c. Display switch.

(1). Controls the brightness of the display pannel.

d. Up/down - link freqcency mode switch.

(1). X1 sets the frequency for the radio set to the down link (RECEIVE) mode.

(2). X2 sets the frequency for the radio set to the up link (TRANSMIT) mode.

e. Sat offset switch.

(1). Presets four different frequencies in the radio set memory.

e. Call switch.

(1). Send mode. Manaly adjusts the desired frequencies up or down.

(2). RCV mode. Set the radio set in the receive/transmit mode.

f. Squelch switch.

(1). Adjusts the squelch to the radio set.

g. XMT power switch.

(1). Manuly adjusts the radio set power output.

SAT COMM. PROCEDURE

1. FUCTION SWITCH TO SAT.

2. DISPLAY SWITCH ON BRIGHT.

3. UP-LINK FREQ MODE SWITCH TO X2.

4. SAT OFFSET SWITCH TO A, B, C, OR D. (FOUR SETS OF FREQS CAN BE PRESET FOR QUICK ACCESS).

5. MODE SWITCH TO XI. (DOWN-LINK FREQ)

6. CALL SWITCH TO SEND.

8. WHILE HOLDING CALL SWITCH IN SEND, ADJUST KHz AND MHz, UP OR DOWN TO DESIRED FREQUENCY

9. RELEASE CALL SWITCH, (AFTER DESIRED FREQS. ARE ENTERED.) THIS WILL LOAD FREQS AND WILL AUTOMATICLY RETURN TO RCV POSTION.

10. ADJUST MODE SWITCH TO X2. (UP-LINK FREQ)

11. REPEAT STEPS 6 THROUGH 9 .

12. RETURN MODE SWITCH TO X2. (TO TRANSMIT)

SPLASH PROCEDURES:

1. PSC-3 VOLUME SWITCH ON FULL.

2. PSC-3 SQUELCH SWITCH TURNED, ON.

3. SELECT PROPER DAY CHANNEL 1 - 6 ON THE KY-57.

4. TURN KY-57 OFF/ON/TD SWITCH - ON.

5. XMT POWER SWITCH ON TH PSC-3, ADJUSTED TO 3/4 POWER. (JUST TOUCHING THE GREEN LINE)

6. SET SPKR/MODE SWITCH, ON KY-57 TO, PT. (PLAIN TEXT.)

7. POINT DMC-120 SATELITE ANTENNA IN PROPER COMPASS DIRECTION AND ELEVATION.

8. KEY HANDSET AND RELEASE. (SPLASH SHOULD BE HEARD), ONCE A SPLASH IS OBTAINED.

9. SET THE SPKR/MODE SWITCH TO CT ON THE KY-57.

10. SET THE VOLUME SWITCH TO THE CENTER ON THE KY-57.

11. SET THE OFF/ON/TD SWITCH TO TD. (TIME DELAY)

12. TURN SQELCH OFF, ON THE PSC-3.

NOTE: HANDSET MUST BE CONNECTED TO THE KY-57 AUDIO CONNECTOR IN THE SECURE VOICE MODE.

LINE OF SIGHT PROCEURES

1. CONNECT PROPER ANTEENA.

2. FUNCTION SWITCH TO LOS MODE.

3. DISPLAY SWITCH ON BRIGHT.

4. UP LINK FREQ MODE SWITCH ON X2. (UHF FREQS ONLY)

5. SAT OFFSET SWITCH SET TO A,B,C, OR D.

6. CALL SWITCH TO SEND.

7. WHILE MANULY HOLDING CALL SWITCH IN THE SEND POSTION, ADJUST KHz AND MHz UP OR DOWN TO THE DESIRED FREQUENCY.

8. RELEASE CALL SWITCH, TO LOAD FREQS. (SWITCH WILL RETURNED TO RCV POSTION)

9. MODE SWITCH TO VOICE. (RADIO IS READY FOR LOS COUMMUNICATIONS)

KY-57 OPERATION:

1. DO STEPS 1 - 9.

2. INSTALL CRYPTO CABLE.

3. SELECT PROPER WEEK CODE.

4. MODE SWITCH TO C. (CRYPTO)

5. ON/OFF/TD SWITCH TO ON POSTION.

6. HANDSET CONNECTED TO KY-57.

7. VOLUME SWITCH 3/4 ON.

VOICE/DATA
BRIGHT
DISPLAY
VOLUME
SQUELCH
SAT OFFSET
FREQUENCY
CALL
QUIET
MODE
FUNCTION
SAT
LOS
SIGNAL

LST-5B

Lesson purpose: To familirize SEAL scout sniper in the operation of the LST-5B.

Objective: To familirize the SEAL scout sniper in the capiblities and operation of the LST-5B for future field use.

LESSEN OBJECTIVE:

a. To familirize the SEAL sniper in:

1. Equipment setup.

2. Equipment characteristics and capabilities and features.

3. Radio set characteristics and capabilities.

4. Description and use of operators controls and indicators.

TRAINING OBJECTIVE:

To Train the SEAL scout sniper in the required skills needed to operate the LST-5B in a field enviroment.

GENERIAL INFORMATION:

1. TURNING ON THE UNIT.

a. Turn on the radio set by turning the volume control knob clockwise.

b. Set the volume control for the desired volume (the SQ control must be in the maximim counter clockwise postion).

c. Adjust the SQ control for the threshold by advancing clockwise slowly, just untill the noise stops and the green "R" light goes out. Advancing the control further will reduce the sensitivity of squelch break.

2. ENTERING PRESET DATA.

1. All preset data is stored in an internal, non-volatile memerory. T he power battery can be changed without disturbing the preset data. when the radio set is turn on the display shows the last display in use when the radio set was turned off, except the BCN and modem will always be turned off at turn on.

NOTE: AFTER ALL PRESETS AND OPERATING MODES HAVE BEEN SELECTED, THE CURSOR SHOULD BE SET OFF THE SCREEN TO PREVENT UNWANTED CHANGES IF THE SET KEY IS INADVERTENTLY PUSHED.

NOTE: THE UNDERSIDE OF THE COVER CARRIES A LABLE INDICATING THE DATE THE MEMORY BATTERY WAS INSTALLED. THE BATTERY SHOULD BE REPLACED EVERY 2 YEARS.

3. STORING PRESET FREQUENCIES.

1. Ten frequenies from 225.000 to 399.995 MHz can be stored in preset channels CH1 to CH9 and in "---" with 5-KHz turning increments.

2. In the frequency/preset display (mode 1a), a solid decimal point indicates the synthesizer is in phase locked. A flashing decimal point indicates that the synthesizer is not in phase-lock and that a fault exists.

3. Frequency cannot be stored when the decimal point is flashing.

4. Improper frequency setting from 200.000 to 224.995 MHz cannot be entered.

5. When storing frequencies in preset channels, you must select the desired preset channel's number before setting the frequency. Preset frequencies are stored as shown in table 2-3.

OPERATING PROCEEDURES.

1. LINE OF SIGHT OPERATIONS.

The followin procedure is for unencrypted operations.

1. ATTACH THE UHF ANTENNA AND THE H-189/GR HANDSET TO THE RADIO.

2. TURN THE RADIO SET ON , VOL CONTROL CLOCKWISE.

3. SET THE SQ CONTROL TO OFF AND SET THE VOL CONTROL UNTILL NOISE IS HEARH IN THE HANDSET.

4. USING THE CONFIGERATION DISPLAY (MODE 2),SELECT THE OPERATING MODES PER TABLE 2-4. THE SCN AND BCN MODES MUST BE OFF.

5. USING THE FREQUENCY/PRESET DISPLAY (MODE 1a), SET THE OPERATING FREQUENY OR SELECT THE PRESET CHANNEL PER TABLE 2-3.

6. TO TRANSMIT KEY THE HANDSET.

7. TO RECEIVE. RELEASE THE PTT SWITCH AND LISTEN TO THE HANDSET.

8. TO ELIMINATE THE BACKGROUND NOISE FROM THE HANDSET WHEN NO SIGNAL IS PRESENT, TURN THE SQ CONTROL CLOCKWISE UNTILL THE GREEN LIGHT GOES OUT. IF THE CONTROL IS SET TO THE FULL CLOCKWISE POSTION, RECEIVED SIGNALS CANNOT BE HEARD.

2. ENCRYPED LOS OPERATION

1. ATTACH THE UHF ANTENNA AND THE TSEC CABLE TO THE RADIO SET.

2. ATTACH THE HANDSET AND THE TSEC CABLE TO THE KY-57.

3. TURN THE RADIO SET ON.

4. USING THE CONFIGERATION DISPLAY (MODE2), SELECT THE OPERATING MODES PER TABLE 2-4. THE SCN AND BCN MODES MUST BE OFF.

5. USING THE FREQUENCY/PRESET DISPLAY (MODE 1a), SET THE OPERATING FREQUENCY OR SELECT THE PRESET CHANNEL PER TABLE 2-3.

6. SET THE SQ CONTROL AS IN STEP 8 ABOVE. RADIO SET IS READY TO TRANSMIT.

3. SATELITE OPERATIONS.

NOTE. OPERATION VIA SATEIIITE REQUIRES THE LST-5 TO TRANSMIT ON ONE FREQUENCY (UPLINK), WHILE IT RECEIVES ON ANOTHER (DOWNLINK). THE LST-5 OPERATES USING SATELLITE CHANNELS WITH BANDWITHS OF EITHER 25 KHz OR 5 KHz. WITH A 25 KHz CHANNEL. THE MODULATION MODE WILL BE FM FOR PLAN TEXT (PT) VOICE AND WHEN USING KY-57 IN THE CT MODE.

NOTE. ON SOME SATELLITE CHANNELS, THE UPLINK POWER LEVEL MAY BE RESTRICTED TO A MAXIMUM LEVEL OF EFFECTIVE, ISOTROPIC, RADIATED. POWER (EIRP). EIRP IS THE COMBINATION OF TRANSMITTER POWER PLUS ANTENNA GAIN MINUS ANY CABLE LOSS THAT MIGHT BE PRESENT

NOTE. EIRP IS GENERALLY EXPRESSED IN dBWs. THE LST-5'S OUTPUT POWER CAN BE ADJUSTED IN 2-WATT STEPS SO AS NOT TO EXCEED THE MAXIMUM EIRP.

1. SAT COMM OPERATING PROCEDURES. (25-KHz)

.

1. ATTACH THE SAT COMM ANTENNA TO RADIO SET.

2. POINT ANTENNA TOWARDS SATELLITE.

3. TURN RADIO SET ON. VOL. CONTROL CLOCKEWISE.

4. USING THE CONFIGURATION DISPLAY (MODE 2), SELECT THE OPERATING MODES(FM, XHI, & CT) PER TABLE 2-4. SCN AND BCN MUST BE OFF.

5. USING THE FREQUENCY/T-R DISPLAY (MODE 1b), SELECT THE PRESET TRANSMIT AND RECEIVE CHANNELS AS SHOWN IN TABLE 2-5.

6. ADJUST THE SQ CONTROL TO A POSTION JUST SUFFICIENT TO TURN OFF THE GREEN LIGHT.

7. CONNECT THE HANDSET THE KY-57 AND THE CABLE LINK BETWEEN THE KY57 AND THE RADIO SET.

8. ADJUST POWER TO THE REQUIRED LEVEL USING TABLE 2-9.

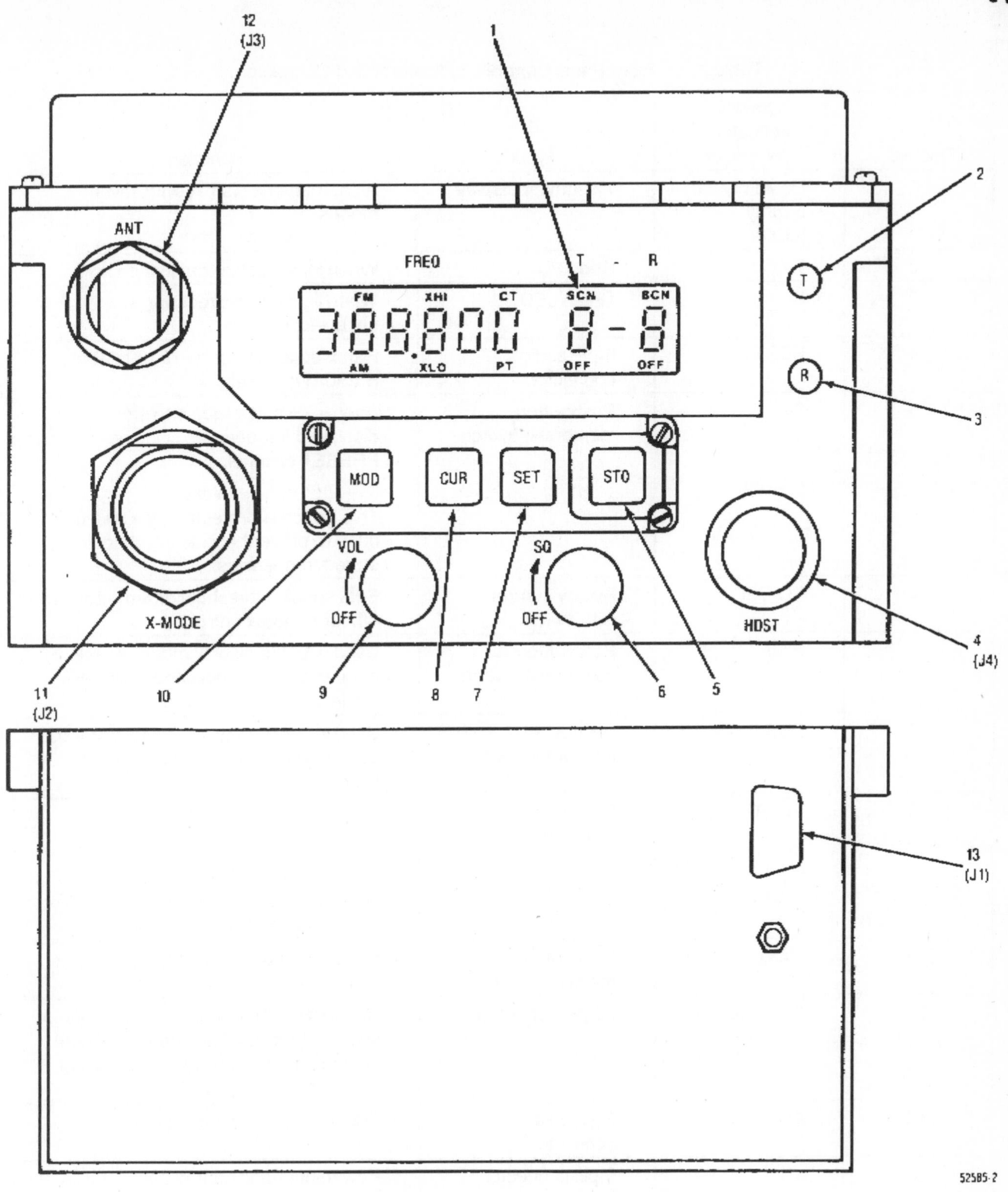

Figure 2-1. Front Panel Controls, Indicators and Connectors

Table 2-1. Front Panel Controls, Indicators, and Connectors

Find No.	Control Indicator, Connector	Type	Function
1	Liquid crystal display (LCD)	7-segment display	Alpha-numeric display with seven modes.
2	T	Red LED	When lit, indicates transmit on.
3	R	Green LED	When lit, indicates receiver is unsquelched.
4	HDST	6-pin audio connector	Handset connector for H-189/GR or H-250/U handset (J4)
5	STO	Pushbutton membrane switch	Used in display Mode 1 to store displayed frequency in selected PRESET channel. ***Note*** This switch is protected by a raised ridge to prevent accidental changes to stored frequencies.
6	SQ/OFF	Rotary control with switch	Sets squelch threshold or turns squelch off, for remote control.
7	SET	Pushbutton membrane switch	Used with the display modes to select frequencies, channels and operating modes.
8	CUR	Pushbutton membrane switch	Used in the display modes to locate the cursor position (indicated by the flashing digit or character).
9	VOL/OFF a) OFF b) VOL	Rotary control with switch	 Full CCW position turns radio off. Continuously variable control adjusts handset audio level.
10	MOD	Pushbutton membrane switch	Selects one of the display/control modes.
11	X-MODE	26-pin connector	Connects radio to peripheral devices such as COMSEC equipment, remote-control unit, test equipment and other radios for retransmit (J2).
12	ANT	N-type RF connector	Connects UHF antenna (J-3).
13	J1	9-pin connector located on back panel	Power input and control connector.

Lesson purpose: To familiarize SEAL scout sniper in the operation of the AN/PRC-113.

Objective: To familiarize the SEAl scout sniper in the capabilities and operation of the AN/PRC-113 for future field use.

LESSON OBJECTIVE:

a. To familiarize the SEAL sniper in:

1. Equipment setup.
2. Equipment characteristics and capabilities and features.
3. Radio set characteristics and capabilities.
4. Description and use of operators control and indicators.

TRAINING OBJECTIVE:

To train the SEAL scout sniper in the required skills needed to operate the AN/PRC-113 in a field environment.

GENERAL INFORMATION. The AN/PRC-113 is a portable, two-band (VHF/UHF) reiecver-transmitter. The radio set is designed for short range tactical ground to air and ground to ground communication. The radoio set can be used with COMSEC TSEC/KY 57 speech security equipment for secure voice communication. Depending on the terrain, ground to ground range varies from one to eight miles. Ground to Air range in smoe cases can exceed fifty miles.

1. RADIO SET CHARACTERRISTICS AND CAPABILITIES.

a. Portable.

b. Two band - automatic band switching.

(1) VHF (1360 channels).

(2) UHF (7000 channels).

2. Tactical mode.

(1) Ground to ground.

(2) Ground to air.

3. Secure voice.

(1) When used with attaching cable and COMSEC/KY-57.

4. Memory.

(1) Will remember its last manuslly selected frequence and all preset channels when turned on from an off condition.

5. Preset channels.

(1) Up to 8 preset channels can be set into radio memory before a mission, or radio can be operated in manal mode.

6. Battery types.

(1) 2-each, non-rechargeable BA-5590/U lithium batteries.

(2) Or 2each, rechargeable BB-590- cadmium batteries.

RADIO SET PRIME FEATURES.

1. DF MODE.

(1) When DF (direction finding) is lit on the display panel the radiom set is transmitting a continuous ton (1000Hz @ 90% modulation). This tone will also be heard in the specker part of the handset.

2. GD MODE.

(1) When GD (guard receiver) is lit on the display panel the radio set will autoomatically receive signals trainsmitted on the guard channel frequency (243.00MHz) and main frequency.

3. SQL MODE.

(1) When SQL (squelch) is lit on the display panel the radio set will provide squelch to the incoming signal. The squelch threshold level is adjustable by the squelch control on the radio set front pannel.

4. LPR MODE.

(1) When LPD (low power transmit) is lit in the display pannel the radio set will transmit in the 2 watt mode. When LPD is not lit, the radio set will transmit in the 10 watt mode.

5. PT MODE.

(1) PT (plain text) will always be lit on the display pannel when the radio set is being operated in the non-secure voice mode.

RADIO SET COMPONENTS.

1. RECEIVER-TRASNSMITTER.

(1) Contains controls, indicators and electronics to operate the set in VHF/UHF tactical modes.

2. BATTERY CASE.

(1) Houses the batteries.

3. BATTERIES.

(1) Supply, power for operation of the radio set.

4. HAND SET.

(1) Provides audio input/output for the radio set.

5. UHF ANTENNA.

(1) Used during UHF operations.

6. VHF ANTENNA.

(1) Used during VHF operations.

7. VHF/UHF ANTENNA.

(1) Adual nahd antenna for VHF/UHF operations.

8. KY-57 BASEBAND CABLE ASSEMBLY.

(1) Connects between RT AUDIO connector and KY-57 unit during secure voice operations.

EQUIPMENT DATA

1. POWER OUTPUT.

(1) 2, or 10 watts.

2. FREQUENCY RANGE.

(1) Receiver/Transmitter. VHF, 116.000 MHz to 149.975 MHz.
UHF, 225.000 MHz to 399.975 MHz.

(2) Guard receiver. (Fixed UHF Frequency) 243.000 MHz.

3. PRESET CHANNELS.

(1) 8.

4. WATERTIGHT.

(1) To a depih of 36 inchs.

5. WEIGHT. 16.7 LBs.

OPERATING INSTRUCTIONS.

GENERAL. Controls, indicators and connectors used by the operator of the radio set are discussed in three groups:

a. KEY BOARD.
b. DISPLAY
c. CONTROLS AND CONNECTORS.

1. KEY BOARD.

(1) The keyboard is used to manually set operating frequencies-set mode of operation-set preset frequency channels.

(2) Keyboard discription and function.

a.

KEY DISCRIPTION	FUNCTION
#1	enters the number one
2	enters the number two
3	enters the number three
4 LPR	enters the number 4 / controls switching between 10 watt output power. LPR will be lit on the DISPLAY when in the 2-watt mode. LPR will NOT be lit when in the 10-watt mode.
5 ACT	5 enters the number 5./ACT this fuction is not used in this configuration.
6 GD	6 enters the number 6/ GD turns the RECEIVER ON or OFF. GD will be lit on the DISPLAY when the guard receiver is ON. GD is not lit when the guard receiver is off.
7 SQL	7 enters the number 7/ SQL turns radio squelch ON or OF SQL is NOT lit when squalch is OFF.

8 TOD	8 enters the number 8/ TOD this function is not used in this configuration.
9 DF	9 enters the number 9/ DF turns on the direction finding tone ON or OFF. DF will be lit on the DISPLAY when DF finding is on. Tone will be heard in the handset. Radio set is continuously transmitting when DF is lit. DF is not transmitting when not lit.
CLR HWT	CLR used to erase an error made while changeing operating frequency./HWT this fuction is not used in this configuration.
0 PST	0 enters the number 0./ PST Turns preset mode ON or OFF. Special mode character P will be lit on the DISPLAY when preset mode is on. P is not lit when preset mode is off.
ENT	The ENT key is used at the begining or completion of operating proceedures to: Stop blinking display Relight battery saver Tune RT to new frequency stored in menory.

DISPLAY.

1. FREQUENCY.

(1) A. blinking number on the display indicates an error made while entering anew frequency.

a. To remove blinking number, first press CLR/HWT key to erase (clear) error, then enter proper number and the erroneouss number will be replaced by the correct number.

2. <u>DISPLAY</u>.

(1) After entring new frequency, all numbers will blink on and off. Press ENT key on KEYBOARD to set new frequency in radio. The display will stop blinking and display the enterd frequency.

3. <u>BATTERY</u>.

(1) The DECIMAL POINT BLINKING indicates low power. This will affect radio operation - replace batteries immediatealy if decimal point is blinking.

4. <u>BATTERY SAVER</u>.

(1) DISPLAY - light will go out within 33 seconds after last function. To re-light the display without changing the frequency or operating mode, press the ENT key. The display will also re-light when any key is pressed.

5. <u>SPECIAL MODE CHARACTERS</u>.

(1) P (PRESET) - Pressing 0/PST on keyboard will show on display if the radio set has been operating with a manual frequency.

a. If the radio set has been operating on a preset channel, the display will show P- and the last preset channel number after preset operation.

EXAMPLE: The display shows - P - 5 , The radio was operating on preset channel 5.

(2) LP (LOAD PRESET) After entering new frequency and while display is blinking, press 0/PST key.

a. The display will show LP- indicating that the load prest mode has been entered into the radio.

b. Press any key 1 through 8, example: 5/ACT Display will show LP-5, Press] key, the display will now show the new frequency entered into the radio memory at preset channel 5.

c. The preset channel for the displayed frequency entered into the radio mem can now be recalled by pressing the 0/PST key. The display will show P-5 , Radio will now receive and transmit on preset channel 5.

6. OPERATING MODES.

(1) DF (DIRECTION FINDING) When 9/DF key is pressed, DF will appear on the display.

a. When DF is lit on the display, a continuous tone is being transmitted by the radio set. This tone is also heard in the speaker part of the operator's handset. The radio set will transmit continuously untill 9/DF key is pressed again. (WARRNING : DONOT DEPRESS 9/DF, KEY UNLESS ANTENNA IS INSTALLED ON RADIO SET).

b. DF Transmite on (1000 Hz @ 90% Modulation).

(2) GD (GUARD RECEIVER) When 6/GD key is pressed, GD will appear on the display.

a. When GD is lit on the display, the radio set - guard receiver will automatically receive messages transmitted by another radio set on the guard channel (243.000 MHz)UHF frequency.

(3) SQL (SQUELCH) When 7/SQL key is pressed, SQL will appear on the display.

a. When SQL is lit on the display, the radio set squelch circuitry is activated. The squelch threshold level is adjustable by the squelch control on the radio set front panel.

(4) LPR (LOW POWER) When 4/LPR key is pressed, LPR will appear on the display.

a. If the radio is not in the LPR mode or if the radio is already in the LPR mode, LPR will disappear from the display.

b. When LPR is lit on the display, the radio set transmitting power is two watts.

c. When the LPR is not lit on the display, the radio set transmitting power is 10 watts.

(5) PT (PLAIN TEXT) PT will always be lit on the display when ever the radio set is operating in the non-secure mode.

OPERATING PROCEDURES

MANUEL OPERATION:

1. CONNECT HANDSET

2. CONNECT ANTENNA (LONGE ANTENNA VHF, SHORT ANTENNA UHF)

3. TURN ON VOLUME

4. PRESS ENT KEY TO ENTER FREQ.

5. PUNCH IN FREQ VHF FREQ BAND 116.000MHz to 149.975MHz
UHF FREQ BAND 225.000MHz to 399.975MHz

6. PUSH ENT KEY TO LOAD FREQ

"NOTE" THE PRC 113 IS READY FOR MANUEL OR ONE CHANNEL OPERATION.

PRESET OPERATION :

"NOTE" PRESET OPERATION IS USED WHEN TWO OR MORE FREQUENCIES ARE USED, UP TO A TOTAL OF 8.

1. CONNECT HAND SET.

2. CONNECT ANTENNA. (LONGE ANTENNA VHF, SHORT ANTENNA UHF)

3. TURN ON VOLUME.

4. DIM DISPLAY SWITCH ON.

5. PRESS ENT KEY TO ENTER FREQ.

6. PUNCH IN FREQ DISIRED, (VHF OR UHF)

7. PUSH PST KEY FOR PRESET CHANNELS 1 through 8. SREEN WILL SHOW P-1 THROUGH 8.

8. PRESS ENT KEY TO LOAD PRESET CHANNEL. THIS MAY BE DONE FOR EIGHT DIFFERENT CHANNELS IN MEMORY.

9. TO RECALL A PRESET CHANNEL PUSH PST KEY THEN THE CHANNEL NUMBER DESIRED , THEN ENT KEY.

NAVAL SPECIAL WARFARE

SCOUT SNIPER SCHOOL

DATE____________________

PLANNING AND PREPARATION OF A SNIPER MISSION

PURPOSE. The purpose of this lesson plan is to ensure that all Special Warfare scout snipers pocess the ability to plan, prepare and carry out a assigned sniper mission.

INTRODUCTION:

All aspects of planning and preparation of a sniper mission are contained in this lesson plan, from the sniper employment officer's responsibilities to the sniper team's responsibilities in planning, preparing, and executing a mission. A sniper patrol is always "tailored" for the mission it is to execute.

1. DEFINITION. A sniper mission is a detachment of one or more sniper teams performing an assigned mission of engaging selected targets and targets of opportunity, and collecting and reporting information, or a combination of these, which contribute to the accomplishment of Naval Special Warfare's mission.

2. SNIPER EMPLOYMENT OFFICER. The responsibilities of the employment officer (usually XO, OPs, Intelligence Officer, SEAL platoon commanders) are:

a. Issuance of necessary orders to the sniper team leader.

b. Coordination.

c. Assignment of patrol missions of employment.

d. Briefing team leaders.

e. Debriefing team leaders.

f. Advising the supported unit commander on the best means to employ and utilize his sniper teams.

g. The most important responsibility.

(1) The sniper employment officer is directly responsible to the commanding officer of what every Special Warfare Team he is attached to for the operational efficiency of his sniper teams.

NOTE: IT IS THE RESPONSIBILITY OF THE SNIPER EMPLOYMENT OFFICER TO FAMILIARIZE HIMSELF WITH THE SNIPER TEAM'S CAPABILITIES AND LIMITATIONS. IT IS THE SNIPER TEAM LEADER'S RESPONSIBILITY TO ENSURE THE SNIPER EMPLOYMENT OFFICER IS WELL ADVISED ON THESE CAPABILITIES AND LIMITATIONS. THE TEAM LEADER IS ALSO RESPONSIBLE FOR MAKING RECOMMENDATIONS AND TO GUIDE THE SNIPER EMPLOYMENT OFFICER ON THE CORRECT METHOD OF EMPLOYMENT OF ALL SNIPER TEAMS UNDER HIS CONTROL.

3. ISSUANCE OF NECESSARY ORDERS TO THE SNIPER TEAM LEADERS. If the sniper employment officer is not available, such as when sniper teams are attached away from their assigned command, the sniper team leader assumes the sniper employment officer's responsibilities. Necessary orders given to the sniper team leader are as follows:

a. Orders providing the sniper team leader with the necessary information, instructions, guidance to enable the sniper team or teams to plan, prepare, and conduct the patrol mission. This information can be given orally and on an informal basis, or as a standard patrol operation order depending on the time available.

b. The responsibility for all detailed planning, when practical, should be given to the sniper team leader. The mission should be described in only the most general terms by the sniper employment officer or the supported unit commander. The routes, targets, locations of firing positions, detailed mission planning, fire support planning and coordination should be the responsibility of the sniper team leader. When he has time, he should prepare and issue, to the observer, or if any other sniper team personnel are attached, a detailed patrol order to ensure that he has planned for every contingency.

4. COORDINATION. Coordination is a continuing, joint effort by the sniper employment officer and sniper teams. The three general areas of coordination are between the:

a. Staff and other staff of other units.

b. Staff and sniper team leaders and units immediately affected by the patrol's operation.

(1) The recommendations for sniper missions to be conducted and the sniper teams to be provided are submitted to the commanding officer for his approval.

(2) The commander may, in his briefing to his staff, inform the sniper employment officer or sniper team leader that snipers may be needed in the overall "big picture."

(3) A sniper patrol is assigned one major mission. The essential tasks required to accomplish the mission are assigned to both the sniper teams and elements of the supporting units.

(4) Whether the sniper mission be a specific mission or a general mission, it must be clearly stated, thoroughly understood, and with the CAPABILITIES of the sniper team.

5. SUPERVISION. Supervision is provided by the sniper employment officer in planning, preparation, and rehearsals, giving the sniper team leaders the benefit of their own training and experience.

6. BRIEFING TEAM LEADERS. Once the commander has stated the need for snipers, the sniper employment officer, if available, must brief the sniper team(s) on the assigned mission.

7. RECEIVING THE ORDER. During the issuance of the order (briefing by the sniper employment officer, united commander or the supported unit commander), the sniper team leader listens carefully to ensure that he clearly understands all information, instructions, and guidance. He takes notes for later planning. After the briefing, he asks questions if points are not clearly understood or not covered.

a. If supporting a SEAL platoon commander, it is the sniper team leader's responsibility to advise the SEAL platoon commander of the proper and optional means of sniper employment to the best accomplish the mission.

8. PATROL STEPS

NOTE: WHEN THE SNIPER TEAM LEADER IS PLANNING AND ORGANIZING THE PATROL, HE SHOULD <u>SOLICIT</u> INPUT FROM THE OTHER SNIPER TEAM MEMBERS TO ENSURE THAT THE OPERATION IS WELL EXECUTED.

THE SNIPER TEAM LEADER WHO DOES NOT SOLICIT INPUT FROM HIS FELLOW OPERATORS IN PLANNING AND ORGANIZING A SNIPER MISSION IS DOOMED TO FAILURE.

a. STUDY THE MISSION

b. PLAN USE OF TIME

c. STUDY TERRAIN AND SITUATION

d. ORGANIZE THE PATROL

e. SELECT MEN, EQUIPMENT, AND WEAPONS

f. ISSUE WARNING ORDER

g. COORDINATE (CONTINUOUS THROUGHOUT)

h. MAKE RECONNAISSANCE

i. COMPLETE DETAILED PLANS

j. ISSUE PATROL ORDER

k. SUPERVISE (AT ALL TIMES), INSPECT, REHEARSE.

l. EXECUTE MISSION.

9. ESTIMATE OF THE SITUATION. In the preparation of the sniper team leader's patrol order, the estimate of the situation is reflexive and continuous by the team leader, upon receipt of his order. Use the following acronym when estimating the situation as if effects the sniper team's employment.

M ISSION
E NEMY
T ERRAIN AND WEATHER

a. STUDY THE MISSION. The sniper team leader carefully studies the mission. Through this, and the study of the terrain and situation, he identifies the essential tasks to be accomplished in executing the mission. (example: Need sniper security support for SEAL squad, day ambush, site grid EJ87659387).

The blocking of routes of escape from a kill zone is an essential task which must be accomplished to execute the mission.

b. PLAN USE OF TIME. Combat situations seldom allow the sniper team leader as much time for planning and preparation as he would like. A well-planned sniper patrol should be planned 24 to 48 hours prior to the time of departure. The sniper team leader should plan his time schedule around specific times (i.e., time of departure, time of attach, etc.) in the operation order.

c. STUDY AND ANALYZE THE TERRAIN AND SITUATION. (Terrain). The sniper team leader and his team study the terrain over which they will be moving, the friendly and enemy situations, and areas of operation.

The sniper team makes a detailed study of maps and aerial photographs (if available) and, if time allows, make a sand table or terrain model of the terrain over which they must pass, to aid in position and route selection. It must include the objective area.

(Situation). The sniper team leader studies the strengths, locations, dispositions, and capabilities of the friendly forces and their fire support that may affect the mission's operation.

The sniper team leader should put himself in the mind of the enemy and come up with an educated guess as to where the enemy is likely to be and what he is likely to do before and after the long-range, percision sniper shot. He should ask himself question about the enemy:

(a) What has the enemy done in the past?

(b) What is he likely to do NOW?

(c) How will the enemy be moving (security activities; patrols, platoons, or companies, etc?)

(d) What will the enemy be trying to accomplish?

(e) What avenues of approach will be utilized?

(f) How will terrain and weather affect his movement?

(g) When will the enemy move?

(h) What is his plan/tactics?

(i) HOW CAN THE SNIPER'S RIFLE AND FIRE SUPPORT PLAN COMBAT LIKELY AND KNOWN ENEMY ACTIVITIES AND CONTRIBUTE TO THE ACCOMPLISHMENT OF THE FRIENDLY SEAL COMMANDER'S MISSION.

d. MAKE A TENTATIVE PLAN. The sniper team leader makes a tentative plan of action. The plan may include:

(a) Type of position.

(b) Location of position

(c) Type of employment

(d) Security backup needs (SEAL squad, etc.)

(e) Target location

(f) Passwords of frontline infantry units

(g) Time of departure

(h) Equipment needed

(i) Route selection

(j) Communications

(k) Call signs and frequencies

(l) Fire support.

NOTE: A tentative plan is later developed into a detailed plan of action.

e. ORGANIZE THE PATROL AND BACKUP TEAM AND SELECT WEAPONS AND EQUIPMENT. If the sniper team is to be inserted as an extension of patrolling activities (by a SEAL security patrol), the security patrol leader maintains operations) and logistic control over the sniper team until the sniper team is dropped off, and then resumes control when the snipers are picked up, on the return of the patrol. (The sniper team leader coordinates with the SEAL patrol leader/backup team on the special equipment necessary for the SEAL patrol members to carry, such as axes, picks, sandbags, ponchos, precut logs, etc., for hide construction, as it may be necessary for the SEAL platoon members to help in the preparation of the hide site.

If the snipers should require immediate aid and extraction, the sniper patrol leader/SEAL backup team commander also coordinate the concept and plan of backup, the normal pickup procedures, and times, if applicable. Both the sniper team leader and the SEAL patrol leader/backup team must be thoroughly familiar with each other's missions, routes, and fire support plans. The SEAL patrol/backup leader must be able to terminate his patrol at any time in order to help extract the sniper team, if necessary. The two leaders must coordinate time schedules as well (i.e., time of rehearsals, time of issue patrol order, time of departure, etc.).

f. COORDINATE. It is the responsibility of the sniper team leader to coordinate with all friendly units. Examples of coordination which must be made are:

(a) Movement in friendly areas. Commanders must be informed of where and when the sniper team will be operating in their sector. Sniper teams must also have information on other friendly activities (patrols) in the area of operations.

(b) Departure and reentry of friendly areas (passwords). Detailed coordination is required here.

(c) Fire support plan and other friendly fires planned in the sniper's area of operations.

(d) Movement of the sniper teams.

g. RECONNAISSANCE. A reconassance may be limited to just a detailed map/ or aerial photograph, or from the point of departure to the limit of sight. Briefings by units who have previusly operated in the area will also be of help.

h. COMPLETE DETAILED PLAN. The sniper team leader ensures that nothing is left out from the predeparture of friendly lines to reentry of friendly lines.

i. ISSUE PATROL ORDER. The way an order is issued is the way it will be received and understood. The order is issued confidently and in a loud and clear voice, continually referring to detailed sandtable of rough terrain sketch.

j. SUPERVISE. The sniper team leader inspects his team and rehearses them.

k. REHEARSE. Visual aids, such as terrain models, blackboards, and sandtables, are used to help ensure COMPLETE understanding by all personnel. If visual aids are not available, planned action are sketched out on paper, sand, dirt, or snow.

An effective method for rehearsal is for the sniper team leader, team members, sniper employment officer, or the supported unit commanders concerned with the mission to talk the entire patrol through each phase of the mission, describing the actions to take place from thetime of departure to return. Terrain models should be used in this method of rehearsal.

k. EXECUTE THE MISSION. The key to effective execution is detailed planning to cover every contingency during the previous patrol steps. What can go wrong, will go wrong. The only defense is a detailed planning. The sniper is always thinking, putting himself in the mind of the enemy, asking himself what would he do if he were in the enemy's shoes.

L. FINAL COMMENT. A sniper ability to accomplish an assigned mission and survive to make it back home, is in direct relation to his acquired sniper skills. This is a never ending process. The sniper who thinks he has acquired all the skills and knowledge relating to being a sniper, and does not seek out to improve his skills, or maintain those skills on a daily basis through formal military schooling and inhouse training, has limited himself and will be limited in his actions in the field.

PATROL ORDER for Sniper Missions

Roll Call

Orientation (with terrain model)

"Hold questions until the end, take notes"

I. SITUATION (As it applies to your mission)

A. Enemy;

1) Weather; PAST, PRESENT, and PREDICTED (for the duration of the patrol)

The following will be listed a minimum of three times:

PMMT	MOONRISE	TEMPERATURE
SUNRISE	MOONSET	HUMIDITY (%)
SUNSET	MOON PHASE/% of ILLUM.	PERCIPITATION/FOG
XENT	WIND (direction/velocity	CLOUD COVER (% or 8ths)

WEATHER AND TERRAIN ARE CONSIDERED FOR THEIR EFFECTS ON VISIBILITY AND TRAFFIC - ABILITY (MOVEMENT)

After listing the weather elements, write a paragraph explaining how the weather will affect you and the enemy. Advantages and disadvantages for both. Information about the weather can be obtained from S-2, S-3, or aviation units.

2) Terrain; Write a paragraph on everything you know about the terrain in the patrol's area of operation. (NOOQA can be used as a basis for this). Explain its effects on both you and the enemy. Advantages and disadvantages. Don't just write about terrain features, include vegetation. Once again - visibility and trafficability.

3) Enemy Situation; Write everything you know about the enemy and everything you can imagine the enemy might do. This will include enemy situaiton, abilities, and probable or predicted course of action. To aid in the arrival of this information use:

P - Size
A - Activities (PAST, PRESENT, PREDICTED)
L - Location(s) of enemy positions and suspected enemy positions
U - Unit(s)/ Uniforms of the Enemy
E - Equipment and weapons the enemy has
E - enemy snipers/counter-snipers
OI - Other Information about the enemy, such as Moral, Resupply Capabilities, NPs, OPs, etc.

D - Defend
R - Reinforce
A - Attack
W - Withdraw
D - Delay

Information about the enemy and terrain can be gotten from the following sources: S-2, S-3, Maps, Aerial Photos, Aerial Recons, Visual Ground Recons, Recon Units, other friendly patrols that have been in the area of operation.

(don't get too carried away.) NASSP
1) Higher - Mission of the NEZT higher unit (Actions and routes too, if applicable)

1) <u>Higher</u> - Mission and location of the <u>next</u> higher unit (actions and routes also, if applicable) If the mission comes from the SEO.SNCO/SrSnTmLdr the next higher unit is Sniper Section. If the mission comes from the SUC (Bn. C.O./Co. C.O.) the next higher unit is Sniper Section if the whole section is attached to the supported unit; the next higher unit is that supported unit (Bn. or Co.) if the whole section is not attached or not assigned a mission as a whole.

2) <u>Adjacent</u> - Mission and routes of units or patrols to the right and left of your area of operation. This is any unit you may have incidental contact with such as: FOs, Recon Units, ANGLICO, Army, etc.

3) <u>Supporting</u> - List any unit that is in general and/or direct support of your patrol, such as: Fire Support - ARTY, NGT, Mortars; Close Air Support; Helo Gunships; Security patrols; Trucks; MedEvac, Insert/Extract aircraft; etc.
Explain in sentences form, then list on call targets as follows:

TGT DESIGNATOR	GRID	DISCRIPTION	REMARKS
The FSCC assigns this. The FSCC will give you a block of TGTS (usually 3). Coordinate with the PO of the Co. you are operating with, to get designators. Check the Bn.s fire support plan, some of your on calls may already be covered.	6 digit coordinate	Where the TGT is, a terrain feature or identifying feature, such as: HILL TOP, ROAD JUNCTION, BRIDGE, ETC.	Examples: Time on Target, Prescheduled Fire, fuse type, type of round, whether the target is on a Check Point, ORD, the objective, etc.

(Units list will be in conjunction with the map overlay.)

Plan as many targets as you want or think you may need. You may not always need as many preplanned, on call targets (designators) as you want, but at least you will have planned for them and know where they are.

4) <u>Security</u> - Missions or locations of security elements in your area of operation. This includes LPs, OPs, and security patrols outside the FTBE.

5) <u>Patrols</u> - The mission and routes of other friendly patrols operating anywhere in your area of operation.

<u>C. Attachments/Detachments:</u>
For a sniper patrol it will be NONE. Never will you have any detachments from your patrol, and 99% of the time you will not have any attachments to your patrol.

II. MISSION
A brief and concise statement that should be written verbatim as received. It contains the 5 W's: Who, What, Where, When, and Why. The why does not have to be included or explained, but you can give the reason for your part in the overall plan. Example: SNTm 3, will move into a position to enable them to reduce key targets, on Bn Objective 'A' lcoated in the vicinity of 356872 to support the Bns. attack on the objective.

III. EXECUTION:

A. Concept of the Operation: A general description or an outline of the conduct of the whole patrol. Given in broad terms from Time of Departure to Debriefing. The concept of the operation is the scheme of maneuver.

B. Specific Tasks NOT IN THE OBJECTIVE AREA:

Covers the actions from the TOD to moving into the ORP, and returning back to friendly lines after executing the mission. The ORP is not in the objective area, but as soon as you step out of the ORP you are in the objective area. Since you can only accomplish, execute, or receive one mission at a time, you will only have one objective area per mission, unless you receive a frag order. A fragmentory order is used to issue supplemental instructions or changes to a current operation order while the operation is in progress. A frag order can also be assigned to a patrol after they accomplish the assigned mission, but before they come back to the friendly area.

These tasks will include, but are not limited to:

1) Escape and Evasion Routes (not in the objective area)

a) Should be easy to find, follow, and utilize.

b) Can use cardinal directions or generalized escape aximuths.

c) Example: Go due north to the second stream and follow that stream to friendly lines.

d) Explain in detail your plan for utilizing your E&E contingency.

2) Security and Actions at Co-ordination Points

a) Security Halts and Actions at Security Halts:

(1) How Long - the team leader will designate how long he will stop for security halts here.

Recommendations: Long security halt - 15 minutes
Short security halt - 5 minutes

(2) When - When will the team leader conduct long or short security halts. Coordination points, enroute.

Where - Security halts should be conducted

(a) Long Halt - before entering Rally Points, Check Points, ORP, Danger Areas

(b) Long Halt - At or before leaving Rally Points, Check Points, the ORP

(c) Long Halt - After crossing Danger Areas, breaking enemy contact

(d) Long Halt - When ever the enemy is sighted or heard

(e) Short Halts should be conducted frequently while enroute

How - How will the security halts be conducted. Should be done in the prone position, feet touching - for signalling, with areas of responsibility for the point man from 9 o'clock to 4 o'clock and for the from 2 o'clock to 10 o'clock.

b) Check Point Grids and Actions at Check Points:

(1) List check points by grid coordinate and number, letter, color, or word codes.

(2) Explain in detail actions to take place before entering, at and before leaving. (Security halts, etc.)

c) Location(s) of Link Up Point(s) and Actions for Link Up:

(1) Grid location for link up (could be a check point of the ORP)

(2) Actions or plans for link up (If applicable, for link ups for friendly patrols, security patrols, partisans, emergency or contingency link ups).

(3) Recommended link up procedure:

(a) Sniper team leader should be in charge of the link up (he is the one being picked up).

(b) Sniper team should be in the link up point prior to directing the patrol in.

(c) Radio communication/voice contact must be maintained at all times.

(d) Have the security (pick-up) patrol halt about 500 meters from the link up point.

(e) Once the security patrol has stopped, have them send 2 - 4 men from the main patrol towards the link up point.

(f) Have the men coming towards the link up point stop every 100 meters to monitor their progress.

(g) Guide the men coming towards the link up point to within 50 to 25 meters of your link up position and have them stop.

(h) At this time, the entire sniper team will get up and go to the men, to the guided back to the main body of the patrol (challenge and passwords.)

d) Location(s) of Release Point(s) and Actions for Release:

(1) A detailed plan for releasing from a security patrol

(a) Sniper team can be released anywhere along the security patrols route.

(b) Sniper team can be released at their ORP (the sniper team leader picks his own ORP).

(c) Sniper team can be released from the patrol after using the security patrol to help build the hide.

(2) Grid location of where the team will be released (a check point can be used).

e) DO NOT INCLUDE RALLY POINTS (they are covered in III.C. Coord. Instr.)

3) <u>Fire Support</u> - List and explain fire support outside of the objective area, such as pre-planned fires to cover routes.

4) <u>Other Tasks</u> - One of the catch-alls. Covers things such as:

(a) Resupply Plan (If not in the Objective Area)

(b) Harbor Site (night pos) Location(s) and Actions at

(1) Locating/ID

(2) Recon

(3) Movement - in and out

(4) Security

(5) Actions with-in

(6) Recommended procedure:

1. Harbor site should located off of the line of march and away from natural lines of drift.

2. The best time to enter the harbor site is near EENT, the best time to exit is near MINT. (Harbor sites can be used in day also.

3. Move to with-in 200 to 100 meters of the tentative harbor site and observe it for approximately 20 to 30 minutes. (This long security halt is also used to observe if you are being followed or to notice any other movement).

4. If the area is all secure, the team leader will leave his gear with the ATL and move into the harbor site (with a .45) to do a recon and to confirm that it is a suitable position.

5. Once the harbor site has been confirmed, the TL will go back to get his gear and lead the ATL into the harbor site.

6. Once the team has moved into the site, one man will emplace the claymores and sensors if they are being used.

7. Once the security of the site has been taken care of, the entire team will remain awake and alert for one hour.

8. After the one hour security waiting period, consider such things as eating, cleaning weapons, head calls, sleeping, etc. Security in the harbor site msut be maintained if these things are to be done. (One man up, one man sleeps). Depending on the situation, you may only want to sleep in the harbor site.

9. Radio communications must be maintained.

10. Approximately one hour before BMWT, the entire patrol should be awake for a long security waiting period of about 45 minutes.

11. The same man who emplaced the claymores/sensors should retrieve them.

(c) Observation Post(s) and Actions are OPs:
   (1) Movement Into/Out of: (from the rear)
   (2) Good field of view (180 degrees if possible)
   (3) Security
   (4) Watches/OP Log

(d) NONE Other tasks simply could be none.

C. <u>Co-ordinating Instructions</u>: This section must be very detailed.

1. <u>Time of Departure (TOD)/Time of Return (TOR)</u>: List the times and dates. This is a commanders control measure.

2. <u>Primary and Alternate Routes</u>: Routes both to the TFFP and back. All <u>LDGS</u> will be magnetic azimuths' and in meter distances. Start from the IRP to check point (this is a leg) to check point, etc. to the ORP and back.

A check point is a control measure. It lets the commander (in the rear) know your location while on patrol or how your patrol is progressing. Check points also aid the patrol in its navigation.

The first check point should be located 400 to 600 meters from the FEBA. (At this check point conduct a long security halt, then one last equipment check. Maintain security and change into ghillie suits if they are to be used on the patrol.)

Check points can also be used as rally points, but are not as close to each other as rally points should be. (Check points should be approximately 300 meters apart. Rally points should be approximately 100 meters

You do not have to physically pass through a check point. For example: A check point used could be a distinguishable hill top that is off the route but you can recognize it easily. (Then call into the rear that you are at check point so and so.)

Plan as many alternate routes as you feel you may need and plan your routes to work together. Do not worry about making an alternate route from the FEBA to the first check point. The alternate route starts from the first check point, do however make an alternate return route to the FEBA.

List your routes. The following is an example:

Primary Route:

From						
From 899759 (IRP)	351°	for 700 M	to	CKPT1	at	897766
" CKPT 1	280°	" 300 M	"	CKPT2	"	895766
" CKPT 2	294°	" 300 M	"	CKPT3	"	892767
" CKPT 3 (ORP)	294°	" 300 M	"	OBJ	"	889678
" OBJ	114°	" 300 M	"	ORP	"	892767
" ORP	394°	" 250 M	"	CKPT 4	"	893764
" CKPT 4	349°	" 300 M	"	CKPT5	"	894761
" CKPT 5	121°	" 600 M	"	IRP	"	899759

Alternate Route: (* This example only has an alternate return route.*)

From						
From ORP	63°	for 850 M	to	CKPT 6	at	899771
" CKPT 6	193°	" 500 M	"	CKPT 7	"	901766
" CKPT 7	194°	" 750 M	"	IRP	"	899759

The routes will be explained in addition to the map overlay.

Principles to Follow for Selecting your route are:

(1) Avoid known or suspected enemy positions and obstacles.
(2) Seek terrain avoiding open areas and offering the most cover and concealment for daylight movement.
(3) Seek terrain permitting quiet movement at night.
(4) Take advantage of the more difficult terrain, such as swamps and dense woods.
(5) Avoid moving along exposed ridges. Move along the slope below the ridge to prevent silhouetting yourself. (Military Crest)
(6) Avoid using trails in guerrilla-infested areas and in the areas between forces who are in contact in conventional operations.
(7) Avoid moving laterally in front of friendly or enemy lines.
(8) Avoid areas which may be mined, boobytrapped, or covered by fire.
(9) Avoid villages, trails leading into villages, wells, and other places where you are likely to meet natives of the area.
(10) Study maps, aerial photos or sketches and memorize your route before you start your mission.
(11) Note distinctive features such as hills, streams, or swamps and their locations in relation to your route.
(12) If possible, "box" your route in with terrain features to aid in navigation.

(13) In unexpected, difficult, or different terrain (and obstacles), such as jungle and swamps, plan an offset (to a known terrain feature.)

(14) Always plan at least one alternate route to use in case you cannot use your primary route.

3. Departure and Re-entry of Friendly Areas:

A) Forward Unit Co-ordination

1) Identify yourself and your patrol with the forward unit commander.

2) State the size (and mission) of your patrol.

3) Give the time and place of departure and return (location of passage points).

4) Give the location of your assembly area and IRP.

5) Tell the commander the area of operation for your patrol.

6) Get the following information from the forward unit commander:
   - a) Terrain and vegetation in front of his sector
   - b) Known or suspected enemy positions
   - c) Recent enemy activities
   - d) Friendly positions - OPs, LPs, Patrols

7) Determine the forward unit's fire and barrier plan

8) Determine what support the forward unit can furnish:
   - a) Guides
   - b) Communications plan between the patrol and the forward unit (to call for support)
   - c) Navigational aids or signals
   - d) Litter teams
   - e) Fire support
   - f) Reactionary squads

9) Exchange call signs and frequencies.

10) Co-ordinate pyro plans, emergency signals, and codes.

11) Confirm challeng(s) and password(s).

12) Ensure the forward unit commander "gets the word" to his personnel (front line, OPs, LPs) of your passage. (You may want to accompany him when he goes to inform the squad leaders, etc.)

13) Ensure the information will be passed on if the forward unit is relieved.

B) Principles for Departure

1) Establish an IRP. The IRP may be occupied or just planned for, but all patrol members must know its location.

2) Security is maintained.

3) Members of the patrol do not move within the friendly units area without a guide to lead them.

4) Final co-ordination is made with the friendly unit commander to ensure no changes have occured since co-ordination was last made. (This may be the first co-ordination made with the friendly unit.)

5) The patrol members will be counted out.

6) The patrol will make a security/listening halt for all members to adjust to the sights, sounds, and smell of the battle area. This halt is normally made beyond the friendly units security positions/Ops/LPs or Final Protective Fires.

C. Techniques for Departure of Friendly Units

1) The patrol arrives at the forward unit and is met by a guide from that unit. The guide will lead the patrol to its IRP.

2) No one, either singly or as a patrol, should move anywhere in the forward units area without a guide.

3) The PL should then make a final co-ordination with the forward unit commander. Here he will learn of any changes that may have taken place since the last co-ordination and of any recent enemy activity that may affect the patrol.

4) Prior to leaving the patrol the PL gives instructions (contingency plan) for what should be done while he is gone. These instructions state:

a) Where he is going (and who he is taking with him)
b) How long he will be gone
c) What to do if he does not return
d) Actions to be taken if there is enemy contact

If all goes well, the PL shouldn't need to re-issue these instructions when he leaves the patrol for final co-ordination.

5) On returning from final co-ordination, the PL may issue a Frag Order to cover any changes.

6) The technique for departing friendly areas depends on the enemy situation.

The common threats are:

a) Ambush and chance contact
b) Indirect fire
c) STANO devices (surveillance, Target Acquisition, Night Observation)

7) Have a security/listening halt after the patrol has moved out of sight and sound of the forward unit (about 400 to 600 meters from the PTFV). This is a short (long) halt to see if the enemy is in the area, and to accustom the patrol to the sights and sounds of the battlefield or the area of operation, before moving on. If a guide is used, have him wait at least ten minutes before going back to friendly lines, after the patrol leaves the first security halt.

8) During a patrol, have frequent security/listening halts to see that the patrol is not being followed and that there is no enemy in the area.

D. Principles of Re-entry of Friendly Areas

1) Ideally, re-entry should be made in the same sector you left from.
2) Establish and occupy a re-entry rally point.
3) Maintain security at the RRP and at the re-entry point.
4) Contact the forward unit for permission to re-enter.
5) Meet a guide at the re-entry point. Normally a pre-coordinated, forward of the front lines password will be used. Have a pre-co-ordinated alternate re-entry signal (pyro).
6) Patrol members will be counted in (to prevent infiltration, especially during reduced visibility.)

E. Techniques for Re-entry of Friendly Units

1) Move the patrol into a rally point near the re-entry point. This RP should be near a prominent terrain feature where the PL can pinpoint his location, with respect to the re-entry point (especially at night.)

2) By radio, alert the forward unit that the patrol is ready to re-enter. Use a code word. The code word must be acknowledged by forward unit before the patrol reconnoiters for the re-entry point. This will indicate that is a guide has been sent to the re-entry point and is waiting, and that security elements, LPs, and OPs have been notified. If you have no communication, do not attempt to re-enter at night. Wait until daylight and use your alternate re-entry signal (pyro plan).

3) If the PL is certain of the re-entry point, he will move forward to make co-ordination.

4) Avoid movement parallel to friendly barriers. If the re-entry point cannot be found, radio higher headquarters and move to another rally point to await daylight or further instructions on the means of re-entry. Do not stay in the same place from which radio transmission was made to avoid RDF.

5) When the re-entry point has been located, the PL will go get the rest of the patrol and bring them to it.

6) The guide leads the patrol through the barriers to a security position previously co-ordinated for debrief.

7) Remember that the re-entry phase of the patrol is one of the most critical.

4. Organization for Movement:

Who will be point, who will be rear during the entire patrol (when, will you switch positions any time during the patrol). List the basic responsibilities for each member of the patrol during movement.

5. Actions at Danger Areas:

A) Five Types of Danger Areas:

1) Linear Danger Areas - roads, trails, fire breaks, streams, rivers, (enemy main line of defense), etc. Both flanks of the patrol are exposed.

2) Small Open Danger Areas - can be hit in one flank and/or the front. (bypass or offset)

3) Large Open Danger Areas - can be hit from anywhere. (Bounds/Over Watch/Leap Frog) (bypass/offset))

4) Series of Danger Areas - similar to large open areas, especially when it's a series of linear danger areas.

5) Danger Areas Within Danger Areas - example: linear danger area within a large open area.

6) (The objective area is a danger area, but is not included in this section).

B) Explain how the patrol will deal with each type of danger area listed above. (Hypothetical Situations) Then list the danger areas along your tentative route, including grid locations, and explain how they will be dealt with.

C) Principles:

1) Avoid danger areas if possible.

2) Plan to offset. RALS rule.

3) Anyone can designate a danger area, the patrol leader determines whether it is or not.

4) The patrol should cross a danger area where observation is restricted, such as a curve in the road, where vegetation comes right up to both sides of the road, or a bend in a stream.

5. Designate a near side rally point and a far side rendezvous point, (if they are not already designated in the patrol order.) The rally point on the near side will usually be the last rally point designated before encountering the danger area. The rendezvous point on the far side will be a safe distance past the danger area near the route of march.

6) Reconnoiter the far side, either by a visual recon from the near side (security halt) or by sending the point man across to conduct a triangular recon. Do not cross until the recon is complete. Triangular recon:

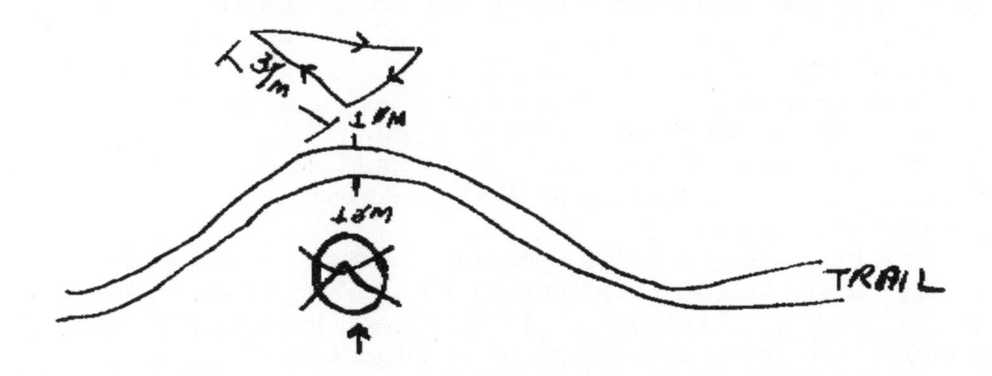

7) If the patrol is split by enemy action while crossing, the man (men) on the far side should go to the rendezvous point and the man (men) on the near side should go to the last designated rally point. The man (men) at the last rally point should then attempt to cross the danger area at a different point, and meet the man (men) at the rendezvous point. If a considerable amount of time has elapsed (or a pre-deisgnated period of time) the patrol should meet at the ORP. (Utilize stealth)

8) Remove evidence that the patrol has crossed, such as footprints.

6. Actions on Enemy Contact:

a) Avoid all enemy contact - that is not advantageous to the mission or is short of the objective area.

b) If contact is made - break contact and either extract the team or continue with the mission.

c) Immediate Action Drills - used when unintentional contact is made and there is no time for giving orders. They must be planned and well rehearsed. Immediate Action Drills are:

(1) Simple - situations calling for IA drills also call for aggressive/violent execution.

(2) Require Speed of Execution - As soon as any member of the patrol recognizes the need for it, he will initiate the appropriate immediate action drill. Instantaneous action usually gives the best chance for success and survival.

Immediate Action Drills are used to:

(1) Counter an ambush.

(2) React to unintentional contact to close range when terrain restricts maneuver. (Break Contact)

(3) Defend against low level air attack.

d) Immediate action should be planned for the following:

1) Air Attack:

a) seek concealment and cover (if you spot A/C and don't think A/C spotted you)

b) lay down perpendicular to the direction of flight and return fire (if you know the A/C has spotted you and is coming in on a gun run) Lead helicopters and prop planes 50 yards, lead jets 200 yards

c) If the aircraft passes you and is going to come in for another run, immediately move and seek better cover and concealment.

2) Chance Contact:

a) Three types:

(1) You see them/they don't see you.

(2) They see you/you don't see them.

(3) You see each other at the same time.

b) Counter measures:

(1) Freeze/Get down/Concealment - Always be ready for contact.

(2) Hasty Ambush - 45 degrees toward the enemy - plan not to initiate the ambush, if the ambush is intiated, immediately leave the area quickly before the enemy can deploy or re-organize.

(3) Break contact - Fire and Maneuver - Clock System, Peel Offs, Receding Leep Frog.

(a) One man fires rapid fire (not full automatic) with the M-16

(b) The other man throws frag/smoke/(WP) grenades

3) Ambush: (With a 2 man sniper patrol, you are dead meat.)

a) Be alert and do not get into an ambush situation.

b) If you are in the kill zone - get out of it immediately. Return fire and break contact.

c) Immediate assault - not feasible.

4) Indirect Fire: If the patrol comes under indirect fire (artillery, mortars, rockets, etc.), the PL will give direction (clock) and distance for the patrol to double-time out of the impact area. Do not seek cover in the impact area, as you will become pinned down. By continuing to move, the patrol is more difficult to hit. (Time permitting, send a SHELLREP).

5) Sniper Fire: The same goes for sniper fire as for indirect fire, GET OUT OF THE AREA. Run in an irratic manner. Once out of the area, call in a SPOT Report (CONTACT) and possibly a fire mission.

6) Minefields/Boobytraps/NBC: (Could be considered Danger Areas)

a) Avoid areas which may contain the above (danger areas).

b) Plan for actions in the event they can not be avoided.

7) HAVE A PLAN FOR ANY CONCEIVABLE SITUATION.

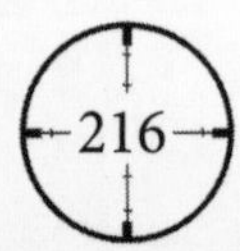

7) Rally Points and Actions at Rally Points:

a) A rally point is a place where a patrol can:
(1) Reassemble and reorganize if dispersed.
(2) Temporarily halt to reorganize and prepare prior to action at the objective.
(3) Temporarily halt to prepare for re-entry of friendly areas.

b) A rally point should:
(1) be easily recognizable.
(2) have cover and concealment.
(3) be defensible for a short period of time.
(4) be away from natural lines of drift.

c) A rally point is physically passed through or passed closely near by.
A rendezvous point is not passed through or passed near by. (tentative)

d) Four types of rally points:
(1) Initial Rally Point (IRP)
(2) Enroute Rally Point (ERP)
(3) Objective Rally Point (ORP)
(4) Reentry Rally Point (RRP)

e) IRP - The IRP is located in a covered and concealed position behind friendly lines. It should be given as a eight digit grid. It is used for a reassembly area if the patrol has been split up. For example, if the friendly unit comes under enemy attack and the patrol is split up, the team will meet at the IRP and man it as a fighting position, or reassemble there after the all clear has been sounded.

f) ERP - ERPs may be planned for tentatively, prior to leaving on the mission (in the patrol order), but are usually deisgnated while enroute by the PL using hand and arm signals. The distance between rally points largely depends on the situation, vegetation, and terrain. Usually rally points are designated every 100 meters. Check points may be used as rally points but this is not advised, as check points are generally farther apart than rally points should be (check points are usually about 300 meters apart).

g) ORP - The ORP is a point near the objective where a patrol makes its final preparations prior to actions at the objective. It can also be used to reassemble, reorganize, and disseminate information after the action at the objective. If possible, the tentative ORP should be out of sight, sound, and small arms range of the objective. For sniper missions the ORP should be 50 to 75 meters from the TFFP. A security halt must be conducted prior to moving in for a recon of the ORP. The recon is conducted prior to occupying the ORP to ensure it is free of enemy and is suitable. Avoid moving in front of the ORP.

h) RRP - The RRP is the rally point located outside the range of small arms fire from the FEBA, approximately 500 meters. If possible, it should be located on or near a prominent (terrain) feature in respect to the re-entry point for ease of locating (the reentry point). Radio contact is made from here to request permission for re-entry.

i) Actions at Rally Points:

(1) If the patrol is dispursed between the FEBA and the first check point/rally point,they should return to the IRP.

(2) If the patrol is dispersed while enroute, patrol members will reeassemble at the last designated rally point. Members will wait a pre-designated time at the RP (covered in the patrol order) before moving on to the previously designated RP and so on back to friendly lines. Before leaving the rally point for the next one, if the patrol has not been reunited, some sort of mark or signal will be left at the rally point. For example: a notch cut in a tree, two feet up from the base.

(3) If enemy activity precludes the use of the last designated rally point, use the previously designated one.

(4) If the patrol is reunited at a rally point, they should contact higher headquarters for instruction on whether to continue with the mission or return to the friendly area.

8. Actions in the Objective Area:

This section must be very detailed. This sub-paragraph covers everything the patrol does from leaving the ORP until returning to the ORP (or the first check point on the return route - if the ORP is not to be used again). The following will be explained in detail:

A) Special Instructions - when utilizing a security patrol

In actuality the security patrol leader is in charge of the patrol until his patrol leaves the sniper team in the hide, BUT he should be very closely advised.

B) Movement from the ORP

(1) Security Halts
(2) Order of March
(3) Responsibilities and Specific Duties
(4) Type of Movement (low) low crawl
(5) Azimuth/Navigation Aids
(6) E & E Plan

C) Tentative Final Firing Position (TFFP) ID and Recon

(1)	Location and Size of the TFFP	10 15 meters in diameter
(2)	How will it be identified	optic gear/NODs
(3)	Who will do the recon	the TL should do the recon
(4)	How will the recon be done	a fan type recon should be conducted
(5)	Where will the gear be staged	the TL does the recon with a .45 pistol only, his weapon (M40A1) and gear will be left with the ATL who is security
(6)	Security (while the recon is conducted)	
(7)	Actions if hit by the enemy	
(8)	Plans for an alternate TFFP	

D) Movement into TFFP/FFP Confirmation

(1) How will the sniper team move into the TFFP (low low crawl - enemy security patrols, LPs, OPs, etc.)
(2) Security plan for moving in
(3) Determine if this is what you are looking for, to accomplish your mission. (What are you looking for - field of view, cover/concealment etc.) Confirm it as your FFP.
(4) Retain communications
(5) Actions on enemy contact/E&E plan

E) Hide Construction

(1) What type of hide will be built
(2) How will the hide be constructed
(3) Who will construct and when
(4) Security plan when building the hide (putting out claymores)
(5) What materials will be used or needed for the hide construction
(6) Possible covering fire to camouflage the noises of construction
(7) Disposal of spoils - where will they go/how
(8) Camouflage
(9) Actions on enemy contact/E&E plan

F) Actions in the Hide

(1) Explain everything you will do in the hide:
 (a) Retain communication
 (b) Fill out range card/Index targets/Sketch area
 (c) Open the observation log
 (d) Update the patrol log
 (e) Modify the fire support plan
 (f) Who will eat/when
 (g) Head calls
 (h) Security in the hide:
  i. watch/observation/radio schedule
  ii. who will emplace/retrieve claymores/sensors
  iii. do not put claymores outside of your field of view
(2) Explain where gear will be located/situated
 (a) where the gear will be - observor on the right, radio in the middle, (with two handsets), hell-box with the team leader, etc.
 (b) a sketch or diagram will accompany the written explanation
(3) Responsibilities and location of each man in the hide
(4) Actions on enemy contact/E&E plan

G) FFP Departure
(1) Who will leave first/order of march
(2) Security plan for departing
(3) Destruction and/or concealment of the hide
(4) Azimuth and distance (return route)
(5) Security halts
(6) Actions on enemy contact/E&E plan

H) Special Instructions in the Objective Area such as:

(1) Link Up Plan (if link up is to be made from the hide)
(2) Counter Sniper Plans
(3) Specific Instructions from Superiors
(4) None

I) Make a contingency plan for every possible situation.

9. Debriefing: When (time and date), Where, Who will and should be present, and the format used for debriefing.

10. Other Actions: Another catch-all. This will cover any actions not specifically detailed elsewhere in the patrol order. For example: Helo-Gunship operations, helo/Aerial resupply, Actions for helo insert/extract, etc.

11. Rehearsals and Inspections:

A) Rehearsals:

(1) Time(s) and location(s)
(2) Who will be present
(3) Order for rehearsals:
(a) Actions in the objective area
(b) Actions at danger areas
(d) Departure and reentry of friendly lines
(e) Organization for movement/Actions at security halts (Actions at check points/Actions at rally points)
(f) Actions in the hide
(g) Other
(4) Rehearse day and night if possible
(5) Rehearse in an area similiar to the area you will be operating in, if possible.
(6) Rehearse in a covered and concealed area - assembly area/IRP.

B) Inspections:

(1) Times and locations (initial and final)
(2) Allow enough time to replace/survey defective gear or correct discrepencies
(3) Sniper teams inspect each others gear
(4) Run about 25 yards to check for noise from gear
(5) Inspect knowledge for mission too, freqs./call-signs, routes, etc.

12. Essential Elements of Information (EEI)/Other Information Requirements(OTR): (could be "none")

A) EEI - critical items of information regarding the enemy and the environment needed by the commander, to relate with other available information, to assist him in reaching a (logical) decision. (observation, log, etc.)

B) OTR - collection of information on other capabilities, vulnerabilities, and characteristics of the area of operations, (which may affect the accomplish-

(5) If wounded is left behind:
(a) make sure he is armed
(b) conceal him
(c) strip him of anything that could identify him as a sniper
(d) record an eight digit grid of his location

(6) Dead:
(a) Med-evac/Extract team - mission is unaccomplishable
(b) Leave behind for later recovery:
(i) strip him of any evidence of being a sniper
(ii) take with you, his equipment, gear, and weapons
(iii) destroy anything you cannot take with you
(iv) cover and bury the corpse
(v) leave one dogtag with the corpse, take the other one with you
(vi) record an eight digit grid for later recovery, a ten digit grid would be better if possible

f) Methods of Handling POWs/Captured Equipment and Documents

(1) POWs
(a) POWs will not be taken. (kill them-quietly) This will be written - as "According to SOP" as killing POWs is a war crime.
(b) If POWs are to be taken either as an OIR or through a frag order - handle them according to the 5 S's:
(1) Search
(2) Silence
(3) Seqregate
(4) Safeguard
(5) Speed

(2) Captured Equipment and Documents - options
(a) Tag as to where and how equipment was obtained and bring it back
(b) Take photos of the equipment if you cannot bring it back
(c) Destroy the equipment
(d) Tag documents and bring them back
(e) Photograph documents

g) Placement of Equipment at FFP

As covered in paragraph III-C-8 (can be discussed again, but does not have to be re-written)

V. Command and Signal

A) Signal

(1) Hand and Arm Signals - verbally explain in writing, show illsutrations, and/or demonstrate. They will include, but are not limited to the following:

a) halt
b) freeze
c) get down
d) move out
e) long security halt
f) danger area
g) ORP
h) rally point
i) O.K.
j) short sec. halt
k) recon (look)
l) check point
m) enemy
n) pace count
o) hasty ambush
p) cannot observe
q) eat

(2) Foot - Tap Signals - (can be used during security halts)

(3) Pyrotechnics Plan - how many and what colors of pop-ups (parachute, star cluster), flares, or smokes will be used for what signal.

(4) Flashlight Signals - with colored lenses and taped cones. (Meta-scopes signals - if they are to be used. Meta-scopes are not recommended anymore because they are infra-red and the Soviets are very big on infra-red detection.)

(5) What is the forward unit's FPF signal - both primary and alternate.

(6) Radio Communications -

(a) List all the frequencies and call signs for the duration of the patrol. These can be obtained from the CECI. Example as follows:

DATE/UNIT	CALL SIGN		FREQUENCIES	
	Primary	Alternate	Primary	Alternate

(b) List the types and times for required reports to be sent. Example: To include but not limited to:

SITREP - every hour on the hour
POSREP - at each check point
SPOTREP - after breaking enemy contact/upon sighting enemy
SHELLREP - as required

Required reports are a commanders control measure and are usually given in the commanders operation order.

(c) Will Dry-Ad/Pele (shackle sheets) be used - ADSIR Authenticate - Down; Shackle - Right

(d) Brevity Codes/Code Words - used on the radio
Examples (not limited only to these):
On the Move - Sleeping
At the ORP - Fishing
At the FFP - Cold Steel
At (a) check Point - A different type of beer can be used for or with the check point number or letter
Alternate Route - Red River

(e) Challenge and Passwords - for the duration of the patrol. They usually change every 12 hours (CEOI).

(1) Front line units (Dep/Reen)
(2) Between other patrols operating in your AO (inter)
(3) For link ups
(4) With-in your own patrol (intra)
(5) Running Password - If you are hit and have to run for the friendly lines and don't have time for, or cannot get proper communications/procedure, or for use in E&E.
(6) The challange and passwords should be used in a sentence (except the running password).

E) Command

1) Chain of Command - not necessary, only 2 of you.

2) Location of PL/APL during halts, etc. - not necessary for the sniper team.

3) Location of the dispatching commander (grid). This will be the SUC, SEO, etc. (COC/CP). Include his frequency and call signs.

ment of (a)/the mission). For example, the sniper team may be assigned an additional duty to the mission, of conducting a route recon (to the objective) either on their way to or from the objective. Or of conducting an "LZ" recon.

13. Annexes: (ranger Hand Book) These are explanations that may be necessary to make the plan more complete. They can be used to promote clarity and understanding areas not covered elsewhere in the order. They follow the normal five paragraph order format and usually are accompanied by a diagram or illustration. They can be used for such things as:

a) Aerial movements
b) Truck movements
c) Aerial resupply
d) Stream crossings
e) Patrol bases (how SnTm will support the PB)
f) Link Ups
g) Small boat movements (IBS)
h) Escape and Evasion

IV. Administration and Logistics

A) BEANS, BULLETS, BANDAGES, AND BADGUYS
B) Part can be copied from the warning order (or "refer to warning order")
C) To include, but not limited to the following:

a) Rations and Water
   (1) How much will be taken
   (2) When and where it will be drawn
   (3) Who will draw it
   (4) When and how you will eat
b) Arms and Ammo
   (1) Who will take what
   (2) When and where it will be drawn
   (3) Who will draw it
   (4) When and where will test firing be conducted
c) Uniform and Equipment (Common to All)
   (1) What will be taken
   (2) How it will be worn
d) Special Equipment
   (1) Who will take what
   (2) How it will be carried
   (3) When and where it will be drawn
   (4) Who will draw it
   (5) When and where will op-checking be conducted
e) Method of Handling Wounded/Dead (friendly)
   (1) First-aid procedures
   (2) Med-Evac procedures/possibilities
   (3) Can walking wounded continue with the mission?
   (4) If wounded is not in danger of death:
      (a) carry him back
      (b) leave behind to go get a friendly security/recovery patrol
      (c) Med-Evac?????

NAVAL SPECIAL WARFARE

SCOUT/SNIPER SCHOOL

DATE_______________

MISSION PLANNING

SAMPLE OPERATION ORDER

1. SITUATIONS

a. Enemy Forces.

(1) Weather. The weather has been hot (90-100°) and himid (80%) for the past week and is expected to continue so for the next week. No rain is predicted during the next 72-hour period. There is a full moon, and visibility will be good. The heat, humidity, and visibility will affect our movement. Here is the astronomical data for 20 July 1986.

BMNT 0520 Moonrise 1730 Temperature 92°
EENT 2100 Moonset 0650 Humidity 63%
Sunrise 0630 Moon phase Full Precipitation 0
Sunset 2020 Wind 3-6 NE Cloud cover 10%

(2) Terrain. The area is full of step fingers and draws, with thick vegetation in the draws near the streams. There exists numerous roads and trails that are not shown on the map. Streams are generally 3 to 4 feet deep and 6 to 10 feet wide, requiring no special equipment for crossing.

(3) Enemy.

(a) There is a battalion-sized enemy unit in the area, but they have been operating in small (8-10 man) groups, with the capability of massing to company size within 24 hours.

(b) Enemy patrols have been reconnoitering our position, possibly indicating offensive action in the near future.

(c) Enemy activity has been spotted in the northwestern section of our AO, with the most recent sighting taking place within 1,000 meters of our lines.

(d) S2 has determined that the enemy is from the 3d Naval Infantry Regiment. The average soldier in this unit is a hard and willing worker and is able to survive and improvise under a wide variety of conditions. He is in excellent physical condition and can bear extraordinary hardship. He has a strong sense of obedience and will attempt to carry out his mission regardless of obstacles or consequences.

(e) There have been contacts within the last 96 hours. All have come during the afternoon hours, with the last coming at 1900 hours last night.

(f) The 3d Regiment is armed with modern Communist bloc weapons and has a mortar and light artillery capability. Enemy uniforms are camouflaged and resemble our own; however, they do not wear body armor. Their sniper/antisniper capability is undetermined.

b. Friendly Forces.

(1) A Co remains in a static linear defensive position with 3d Platoon in reserve.

(2) 1st Platoon is on the left flank, and 2d Platoon is on the right flank. A Co is bounded on the left by B Co, with C Co to our rear in reserve.

(3) We have one 81-mm mortar section in direct support and G Battery in general support of the mission with the following targets plotted:

TGT NO.	LOCATION	DESCRIPTION	REMARKS
AW1001	897766	Hill/Checkpoint #1	HE (81)
AW1002	895766	Hill/Checkpoint #2	HE (81)
AW1003	892777	Finger/ORP	HE, WP (81)
AW1001	899799	Hill/OBJ	HE, VT (155)
AB1002	901790	Hill/Checkpoint #3	HE, HC (155)
AB1003	889791	Finger/Checkpoint #4	HE, HC (155)
		(Alternate route)	
AW1004	891808	Hill/Checkpoint #5	HE (81)
AW1005	901805	Hill/Checkpoint #6	HE, WP (81)
AW1004	891791	Draw/Checkpoint #7	HE, VT (155)

(4) There will no other patrols operating in the company area of operations.

c. Attachments and Detachments. None.

2. MISSION

Sniper team (#1) will depart at 1900 hours 20 July 1986 and proceed to the vicinity of grid 892777 to establish an OP. Collect and report all relative information and modify A Co fire support plan accordingly.

3. EXECUTION

a. Concept of the Operation of the Sniper Team in the Objective Area. This is a brief statement of the "whole picture," or an outline of the conduct of the mission, followed by a detailed report of all the actions the sniper team will accomplish from the ORP to the FFP and back to the ORP, to include movement, routes, security halts, pacing, selection of FFP, construction of FFP, placement and operation of logistics, individuals' duties in the objective area, objective location, identification and engagement, reports, immediate action drills, contingency plans and actions, air support/fire support employment, and a host of others.

b. Actions of the Sniper Team Not in the Objective Area. A detailed report of all tasks or actions accomplished outside the objective area, i.e., from the IRP to the ORP (including action in the ORP) and back to the IRP. Actions may be similar to 3a, but in 3b it is necessary to include departure/ reentry procedures, actions at rallying points, etc.

c. Coordinating Instructions. Depending on the requirements of the instructional situation, the student may be required to duplicate information previously noted in 3a or 3b. If not previously written, then all required information of the following matter will always be covered in detail: infiltration plan (in annex __); linkup plan, annex ___; TOD; TOR; primary and alternate routes; departure and reentry of friendly lines; organization for movement; action on enemy contact; rallying points and actions at rallying points; actions in the objective area; debriefing; other actions; rehearsals; and inspections.

The following is an example of a detailed 3c (Coordinating Instructions).

(1) Infiltration plan. Annex A.

(2) Linkup plan. Annex B.

(3) TOD. 2100/9 Jun 86.
TOR. 06--/12 Jun 86.

(4) Our primary route will be as follows:

From 899759 (IRP)	351° for 200 m to Checkpoint #1 at 897766
Checkpoint #1	280° for 300 m to Checkpoint #2 at 895766
Checkpoint #2	294° for 300 m to Checkpoint #3 OBJ at 889768
OBJ	114° for 300 m to ORP at 892767
ORP	349° for 250 m to Checkpoint #4 at 893764
Checkpoint #4	349° for 300 m to Checkpoint #5 at 894761
Checkpoint #5	121° for 600 m to IRP at 899759

Our alternate route for return from the objective is as follows:

From ORP	63° for 800 m to Checkpoint #6 at 899771
Checkpoint #6	193° for 500 m to Checkpoint #7 at 901766
Checkpoint #7	194° for 750 m to IRP

All azimuths given are magnetic.

(5) Departure and reentry of friendly areas. We will move out from our assembly area to a covered and concealed position short of the FEBA. This will be our IRP, at grid 899759. We will do one last personal gear inspection and communication check.

I will move forward and make final liaison with the 2d Platoon leader. I will request the latest information on the enemy, the terrain to the front, known obstacles, and locations of OPs/LPs. I will check to ensure there has been no change in communication procedures or the fire support the unit can provide. Additionally, I will confirm the challenge/password and determine if the same men from 2d Platoon will be manning the position

upon our return. If not, I will ensure that they notify their relief of our expected return. I will pick up the guide they have provided to take us through the wire and minefield and then verbally call for you to move up. The patrol will then depart through the 1st Squad, 2d Platoon sector in single file, with the guide at the point, you second, and myself in the rear. The 1st Squad leader will count us out and give me the count as I pass his position. When we reach 200 meters forward of friendly lines on a magnetic azimuth of 351°, we will stop and execute a long security halt. After determining that all is secure, sniper patrol 1 will continue on a magnetic azimuth of 351° with you at point and myself in the rear. The 2d Platoon's guide will wait 10 minutes after our departure before returning to friendly lines.

For reentry, we will halt approximately 500 meters short of friendly lines while I contact 2d Platoon for permission to reenter. I'll report our position and wait for the "OK." While waiting we will maintain 360° security, with emphasis to our rear 180°, which will be your responsibility. 2d Platoon will then contact all OPs/LPs, informing them of our return. Upon 2d Platoon's permission we will move forward to a position 200 meters outside of friendly lines, and I will again contact 2d Platoon, inform them we are within local security, and await final permission to reenter.

When we receive the OK, we will move forward, me first, you second, and make contact with forward friendly lines, utilizing the appropriate challenge/password procedure. I will enter friendly lines first and count you in. The alternate reentry signal is the firing of a double white star cluster before moving forward.

(6) Organization for movement. At all times during the patrol (with the exception of the noted reentry procedures) you will lead, navigating and providing frontal security. I will follow, pacing and providing security to the rear.

(7) Actions at danger areas. We will avoid danger areas if possible. Small open areas will be negotiated as the situation dictates. I will determine the method at that time. (At this time the patrol leader would mention any known danger areas, their location, and method of avoidance or navigation.)

(8) Actions on enemy contact. Our patrol is offensive in nature; however, we will avoid all enemy contact in situations not advantageous to us or short of our objective area.

(a) Chance contact. If we see the enemy and he has not seen us, we will quickly assume a concealed position and prepare ourselves for contact but remain silent and allow the enemy to pass. If we are seen by the enemy, you will engage them with rapid fire (not full-auto) with the M-16, and I will engage them with fragmentation and smoke grenades. Under cover of smoke, I will yell "MOVE," followed by a direction. We will then move to our last rallying point or checkpoint, whichever is closest.

(b) Break contact. Our withdrawal will be by the clock method or fire and maneuver, depending on the effectiveness of fire. Whenever possible, our withdrawal will be covered by smoke and/or fire support.

(c) Air attack. We will move laterally out of the aircraft's direction of approach and place our bodies perpendicular to the aircraft's direction. If it returns, or more than one aircraft attacks, we will disperse and see adequate cover.

(d) Artillery/mortar attack. We will disperse and seek adequate cover, attempting to determine the direction of hostile fire and the size of ordnance.

(9) Rallying points and actions at rallying points. The IRP will be at 902757. I plan to use checkpoints as RPs (en route) and will designate them by hand and arm signals. In the event we are dispersed by incoming fire while behind friendly lines, we will reorganize at the IRP 10 minutes after the all clear is sounded. If we are dispersed after leaving friendly lines but before the first RP, we will return to the IRP. Should the patrol become dispersed for any reason, or should an individual become lost, we will return to the last designated RP and wait 45 minutes. If I have not returned in that period, call HQ and request instructions. If communication is not established, use 220° as an escape azimuth to the hard-surfaced road and return to friendly lines. Utilize challenge/password procedures at all times at rallying points. If either man departs the rallying point alone, he must leave a knife mark two feet up on a tree indicating he was there.

(10) Actions in the objective area. Same as 3a.

(11) Debriefing. The patrol will be debriefed immediately upon return at S2, utilizing the NATO patrol report form. The S2, the S3, and the commander will be present.

(12) Other actions. (here you would cover any actions not specifically detailed elsewhere in the patrol order--resupply, emergency extraction by helicopter, etc.)

(13) Rehearsals and inspections. Rehearsals will be in the following order: actions in the objective area, actions on enemy contact, departure and reentry of friendly areas, and organization for movement, to include long and short security halts. The initial inspection will be conducted immediately after the patrol order, followed by rehearsals, surveying of any defective gear, chow, and then the final inspection. All rehearsals and inspections will be conducted in the assembly area.

4. ADMINISTRATION AND LOGISTICS

a. Rations. No change from WO.

b. Arms and Ammo. No change from WO.

c. Uniforms and Equipment. No change from WO.

d. Method of Handling Wounded and PWs. Wounded will continue on the mission, if possible, after first aid treatment. If it is impossible for them to continue, the team leader will determine if they should be MEDEVAC'ed. The dead will be buried and noted for future pickup. PWs will be handled according to sniper SOP.

e. Placement of Equipment at FFP. No change from para 3a.

%. COMMAND AND SIGNAL

a. Signals.

(1) Signals in the objective area during departure and reentry of friendly lines and movement will be as discussed earlier and as we will rehearse.

We will use the following arm and hand signals:

Halt	Long security halt	Short security halt
Freeze	Danger area	Recon (look)
Get up	ORP	Checkpoint
Move out	OK	Pace count
Enemy	Rallying point	

NOTE: A picture of each signal should be given to each individual.

(2) Communications with higher headquarters.

(a) The following call signs and frequencies will be good for the duration of the patrol:

UNIT	CALL SIGN	FREQ P	FREQ A
Company	AB6	30.15	41.75
2d Platoon	AF7	30.15	41.75
1st Platoon	PC3	30.15	41.75
ST #1	HB1	30.15	41.75
FDC 81	AP1	32.45	48.50
FDC 155	HE6A	34.65	44.50
TAC AIR	LTA4	39.50	49.65

(b) The following reports are required:

SITREP--every hour on the hour.
POSREP--at each checkpoint.
SPOTREP--after enemy contact.

(c) Additionally, departure and reentry of friendly lines will be reported to A co. We will use the following code words:

BLUE STREAM at ORP
COLD STEEL alternate frequency
RED RIVER alternate route
JADE BOY at FFP

(d) The challenge and password are as follows:

The challenge will be SEPTEMBER
The password will be HEAVY LADY

b. Command.

(1) Chain of command. No change from WO.

(2) Location of leaders. During movement I will bebehind you. A Co. commander will remain with A Co HQ.

POST-MISSION NONPERMISSIVE DEBRIEFING FORMAT

1. ADMINISTATION:

a. Name of debriefer.

b. Name of individual being debreifed.

2. ORGANIZATION:

a. COMPOSTION OF THE ELEMENT.

b. Postion within the element.

c. Other menbers of the element.

3. MISSION:

a. Primary mission assigned to the element.

b. Additional mission.

4. EEI/OIR/ PIR (PRIMARY INTEL REQUIRMENTS).

5. FRIENDLY FORCES:

a. K.I.A.

b. W.I.A.

c. P.O.W.

d. MISSING

6. SUMMARY OF ACTIVITIES:

A chronological, detailed statement emphasizing time, movement activities, and observations within the area of operations.

a. Infiltration (time and place)

b. Movement (direction and distance)

c. Observation of human activity

(1) Where were people seen.

(2) When.

(3) Number.

(4) Civilian or miltary.

(5) Ethnic group, language, etc.

(6) Clothing (color, condistion). Footgear, headgear, trouser, shirts.

(7) Equipment (color, size, shape, condition).

(8) Small arms (condition and type)

(9) What were the people doing?

(10) If miltary, well-disciplined or para-miltary.

(11) Apparent physical condition.

d. Observation of structures:

(1) Where located.

(2) How many.

(3) Shape, size, purpose.

(4) Construction materials.

(5) Markings.

(6) Contents of structure.

(7) Estimate of last use.

(8) Indications of family occupancy.

(9) Animals or animals pen near structures.

(10) Crops close to structure.

e. Oservation of animals.

(1) What type, how many.

(2) Wild or tame.

(3) Condition.

(4) Drayage animals.

f. Actions at the objective:

g. Exfiltration:

7. ENEMY FORCES:

a. Enemy encountered during i filtration (SALUTES).
SIZE, ACTIONS, LOCTION, UNIFORM, TIME, EQUIPMENT, SNIPERS.

b. Enemy encountered during movement to the objective (SALUTES).

c. Enemy objective.

(1) Guard force (SALUTES).

(2) Emplacements:

(a) SALUTES.

(b) Fields of view.

(c) Fields of fire.

(d) Stage of development.

(e) Guards/personnelat each.

(3) Automatic weapons:

(a) SALUTES.

(d) Type of weapons.

(4) Crew served weapons:

(a) SALUTES.

(b) Fields of view.

(c) fields of fire.

(d) Direction of weapons orientation.

(e) Amount of ammunition.

(5) Apparent state of readiness.

(6) Missing periods and procedures.

(7) Night vision equipment:

(a) Number.

(b) Individual.

(c) Attached to weapons.

(8) Resupply procedures:

(a) When.

(b) How.

(c) From where.

(9) OPs & LPs.

(a) SALUTES.

(b) Communications procedures.

(10) Early warning devices:

(a) Location.

(b) Type.

(c) How activated.

(11) Communications equipment:

(a) Type.

(b) Location.

(12) Use of aviation support.

(a) Rotary wing (SALUTES).

(b) Fixed wing (SALUTE).

(c) Fast movers (SALUTE).

(d) Armament.

(13) Enemy tactics:

(a) What was the enemy reaction to them?

(b) How did the enemy indicate that he was aware of the teams presence in the

(c) Was the team followed? - By how large of force?

(e) Was the team surrounded?

(f) Did the enemy attempt to avoid contact?

(g) What reaction did the enemy have when he was attacked?

(h) 'What action did the enemy have when helicopters arrived to remove the tea insert a large force?

(i) What signal were used?

(j) Discipline of eneny force?

(k) Indications of enemy training?

(14) Mines:

(a) Exact location.

(b) Details of placement.

(C) Number of mines.

(e) Detonation of mines and results if known.

8. TERRAIN:

a. Landform.

b. Vegetation:

(1) Lowland.

(2) Ridge and mountains sides.

(3) High ground, ridge tops and hilltops.

c. Rivers and streams:

(1) Location.

(2) Width.

(3) Depth.

(4) Current (speed and direction).

(5) Slope and bank.

(6) Composition of soil on bottom and banks.

(7) Dimensions of dry bed.

(8) Are large streams navigable?

d. Trails (Identify on map).

(1) Direction and loction.

(2) Width.

(3) Estimate of use (man or animals, footprints describe prints:)

(a) Barefoot.

(b) Cleated soles.

(c) Hard soles.

(d) Direction of movement.

(4) Overhead canopy.

(5) Undergrowth along sides of trail.

(6) Direction signs, symbols, signals found along the route.

(7) Surface characteristics (hard packed or soft earth, dead vegetation; light brush growth, etc.

e. Roads:

(1) Direction.

(2) Width.

(3) Surface material.

(4) Indications of movement.

(5) Maintance of road (craters repaired, etc.).

(6) Description of vehicals tracks.

(7) Chokepoints.

(8) Type vehicles if observed.

f. Bridges:

(1) Type construction and description.

(2) Capacity.

(3) Number of lanes.

(4) Width and length.

g. Soil:

(1) Appearance (color).

(2) Hardness (dry, wet, muddy, very muddy).

(3) Standing water.

h. Note deviation from map of landforms, treelines, waterways, trails, etc.

9. LZ's, PZ's, DZ's:

a. Location.

b. Reference point.

c. Description and demension.

d. Open quadrant.

e. Recommended track.

f. Obstacles.

G. Additional information.

10. Weather:

a Visibility.

b. Cloud cover.

c. Precipitation.

d. Ground fog.

e. Winds.

f. Temperatures.

g. Illumination.

h. Effect on personnel.

11. Communications:

a. Was jamming encountered?

b. Problems in contacting air-relay.

c. Difficulties with radio set.

d. Indication of enemy RDF capability.

e. Was any ground relays used, if so problems encountered, if any.

12. AIR-STRIKES:

a. How many were called?

b. Locations.

c. What results.

d. Was the ordnance effective against the target?

e. Include those not called by team but observed in area.

f. Effects of other types of air support.

13. ADDITIONAL INFORMATION:

a. Anything not otherwise covered.

b. General estimate of the extent of military activity in the area.

c. Signals.

(1) Was there an identifiable pattern to the signals? What was the pattern? Are different methods of signalling integrated in the system?

(2) Were the signals related to enemy activity?

(3) What was the apparent meaning of the signals?

(4) Are different types of signals used in different areas?

14. RECOMMENDATIONS:

a. Items of equipment or material that can be used to improve our operational capability.

RANGE BRIEF

1. Shooters will be devided into two groups:

   a. Shooters (Element one)

   b. Pit detail (Element two)

2. Senior man will ensure all his people are mustered with the required equipment:

   (1) Personnel equipment.

   a. Weapon

   b. 3 magazines

   c. Ammuniton - 100 rds

   d. Field uniform

   e. Boots

   f. Pencel or pen

   g. Ear protection

   h. Rifle sling

   i. Water
   j. Logbook
   k. Spotting scope

   l. Tripod
   m. Rain gear

   (2) Team equipment

   a. 2 cans spray glue.

   b. 1 bottal of white out for marking sights

   c. 5 300 R for communictions

   d. 2 plastic trash bags

2. Range check out procedures.

   (1) Range safty officer will check the range out

(2) Team leader / LPO will muster his element on the 200 yard line prior to shooting, so any last mimute word can be pasted by the range safety officer.

a. Once the word has been pasted the elements will proceed to their asigned postions . (Pits or Firing line)

3. Firing line procedures.

(1) The range safety officer will be incharge of all safety and range operation

(2) The senior man of element one will be incharge of mustering his people at their asigned postion. (Firing line) his responibilities will be:

a. Ensure that there is enought ammunition to complete the days evolution for both elements, staged on the 200 yard line.

b. That all 300 R hand raidos (5 each) are stage on the firing line prior to the days evolution.

c. That both safety range flags are up on the flanks of the 500 yard line.(ran safety flags are stored at the U.S.M.C. Range shack on the 500 yard line)prior to the start of the days evolution.

d. Serior man will also designate one man for range guard prior to mustering on the 200 yard line prior to the days evolution.

e. Ensure that there are two large plastic bags staged at the 200 yard line pri to the days evolution for trash and brass colection.

f. That all brasscaseing are policed up at the end of the days shooting evolution.

4. Pit detail.

(1) The senior man of the second element will be incharge of running and supervizing the pit detail. His responiblities will be:

a. Assign a designated range guard prior to the days evolution.

b. Ensure that the range keys are cheched out to unlock the target shed. (Keys are located at the U.S.M.C. range shack.)

c. Ensure that two fresh cans of spray glue are taking down to the pits prior to the start of shooting evolution.

d. Ensure that a 300 R hand radio is taken down to the pits prior to the start of the shooting evolution for comms. back to the firing line.

e. Ensure that both range guards are issued range safety flags prior to the range guards departing to their assigned postions. (Range safety flags are stored in the target shed in the pits).

f. Ensure that all targets used on the days shooting evolution are in a good state of repair, prior to puting the targets in their frames.

g. Prior to the days shooting evolution, fresh target faces will be pasted on the target frames. (Repair target faces are located in the back room of the target shed)

h. Prior to the days shooting evolution 200 yard targets will be pasted on the target frames , and after each pit change.

i. That each man is issued a range box to mark targets with (range boxes are located in the back room of the target shed) the Team leader will ensure that the following equipment is in the range boxes:

1. Black and white target pasties to mark bullet holes - 2 - rolls ea.

2. 3 inch spoter spindals - 2 - ea.

3. 1 inch spoter spindals - 10 - ea.

4. Target tie ins - 2 - ea.

j. Pit O.I.C. will be incharge of keeping time, for running targets into the air, and running them back into the pits.

k. Responible for conducting a good police call at the end of the days shooting evolution, and that all targets, range boxes and range guard flags are secured prior to locking the target shed.

5. Range guards.

(1) Normally there are two range guards on each flanks.

(2) Responibilities:

a. Keep beach security of the impact area to enclude:

1. Keep all personnel out of the impact area.

2. Ensure that there is no boat triffic in the impact area, out to a range of 2 miles and 45 degrees to the flanks.

3. Ensure that there is no low flying aircraft in the inpact area.

NOTE:
If any of these seduations occur, radio comms back to the firing line must be made for a immediate check fire, until the inpact area is clear of all hazards.

4. Equipment needed.

1. 300 R hand raido.

2. Range safety flag.

NOTE:
Range guards willnot leave their postions unless releaved by the range safety officer.

6. Voice commands. (Naval qual. course)

NOTE: the following voice commands will be given by the range safety officer only, shooters will not load and lock any weapons unless directed by the range safety officer.

(1) Loading and locking of weapons.

a. Shooter stand, with a magazine and 5 rounds, load and lock, all ready on the left, all ready on the right,all ready on the firing line, shooters when your targets appear you may commence firing.

(2) On the command from the firing line range safety officer, to the pits O.I.C., all targets will be rasied into the air for what ever time that is requested.

a. Sighters - 2 mins. for 2 rounds.

b. Slow fire - 10 mins. for 10 rounds.

c. Rapited fire - 60 sec. for 10 rounds.

(3) The pit O.I.C. will keep the time , for running all targets in the air and lowing them back into the pits aftr the designated time has lapesed.

7. Running and markinng of targets.

(1) Sighter rounds.

a. 2 rounds for 2 mimutes.

b. 2 sighter rounds will be given prior to the start of the naval qual course.

c. They can be taken in any shooting postion.

d. Targets will be rasied, lowerd, and marked for each round fired.

e. Sighter rounds donot count for score.

(2) Slow fire.

a. Slow fire will be for 10 rounds for 10 minutes.

b. Slow fire will be conducted from the following postions:

1. Standing.

2. Prone.

3. Setting.

c. Targets will be pull, marked, recorded for score,and rasied to show the shooter the inpact of his round, after which the targets will remain in the air untill that target receives a hit from another round.this procedure will be done for the entirer 10 rounds.

(3) Rapit fire.

a. Rapit fire will be for io rounds for 60 seconds.

b. Rapit fir will be conducted from the following postions.

1. Setting.

2. Prone.

c. Rapid fire will be for 10 rds. for 60 sec., with a magazine change.

d. The shooter will have two magazine with five rounds each.

e. On rapit fire the pits will run the targets up, on the command from the range safety officer on the firing line. The pit O.I.C. will rasied all the targets in the air at once, for 60 seconds, the pit O.I.C. will keep the time to keep the targets in the air. After 60 seconds has lasped, the pit O.I.C. will give the command to lower all targets back into the pits, after which all targets will be marked, recorded for score, and rasied back into the air, to show the shooter the inpact of his rounds.

8. SAFETY.

1. All weapons not on the firing line will be on safe, and unloaded.

2. Any accidental discharges of a weapon will be grounds for evaluating your assignment at this command.

3. No one will be allowed to go downrange without the permission of the range safety officer.

MARKSMANSHIP TEST

1. THE PURPOSE OF THE MARKSMANSHIP TEST IS TO EVALUTATE THE STUDENT ABILITY TO ENGAGE TARGETS AT VARIOUS RANGES, SCORING ONE POINT PER HIT WITH 80% ACCURACY.

2. The student will be required to engage stationary targets at ranges from 300 to 1000 yards, moving targets 300 t0 600 yards and pop up targets from 300 to 800 yards and must get at lease 80% of the total rounda fired.

3. Neeed equipment.

a. Communications equipment .

b. Score cards for the pits and the line (Only the pit score is valid.) Verifiers should be present.

c. Range safty officer.

d. Corpsman.

e. Emergency vehicle.

f. 1000 yard known distance range.

3. CONDUCT OF ENGAGING STATIONARY TARGETS FROM 300 TO 1000 YARDS.

a. Each team will be assigned a block of eight targets, each block of which will be designated with the left and right limits marked with a 6-foot target mounted in two respective carriages. Thus, the right limit for one block will also serve as the left limit of the next block. The following targets will serve as left and right limits respectively: 1, 8, 15, 22, 29, 36, and 43. The stationary target will be mounted in the left limit target carriage of each block.

b. The first stage of fire at each yard line(300,500,600,700,800,900 and 1000) will be stationary targets from the supported pron postion. Command will be given from the center of the line to load one round. The sniper and partner will have three mimutes to judge wind, light conditions, proper elevation hold, and fire three rounds with the target being pulled and marked after each shot. After the three minute time limit has expired, all sattionary targets will be pulled down, cleared, and will remain in the pits. There will not be a change over between sniper and observer untill the sniper has engaged his moving, and popup targets, which should begin immediately after pulling the stationary targets in the pits.

2. CONDUCT OF ENGAGING MOVING TARGETS FROM 300 TO 600 YARDS.

a. Each student will remain at their respective firing point after engaging stationary targets, so they can engage their moving and popup targets, within the assigned block of eight targets. One of the butt pullers will postion himself at the left limit with the moving target, ready to move when the stationary stage is completed.

b. The second stage of fire at each yard line (300, 500, 600,) will be moving targets. The command will be given from the center of the line to load two rounds. Once the entire line is ready, a moving target will appear on the left limit of each block of targets, moving left to right. The sniper and partner will have approximately 15 to 20 seconds(the amount of time it takes the student to walk from the left limit to the right limit) inwhich to fire one round. The next target will move from the right limit to the left and again the sniper and his partner will have 15 to 20 seconds to shoot one round. The target will be run up after each hit. It will also be up to the partner to advase the sniper on where his rounds are impacting (high, low, left, right).

3. CONDUCT OF ENGAGING POPUP TARGETS FROM 300 to 800 YARDS.

a. The next stage of fire will be popup at each yard line (300,500,600,700,800). Each student will remain at their respective firing point after engaging moving targets, so they can so they can engage their popup targets, within the assigned block of eight targets.

b. The command will be given from the center line to load two rounds. once the entire line is ready, a popup target will appear for agiven amount of time depending on what yard line the shooter is on will determine the amount of time the popup target will appear and the size the target will be:

(1) Time formala - yardline x 1 = seconds of exsposure. (Exsample 300 yardline
3 x 1= 3seconds).

(2) Target size formala - yardline x 2 minutes of angle (Exsample 400 yardline
4 x 2 = 8 inches in target diameter).

the shooter will engage two popup target on each yard line while his partner calls wind and recored all information in his data book.

c. Popup targets willnot be engaged past 800 yards. Therefore, five rounds will be fired and scored on stationary targets at 900 and 1000 yards.

4. TEST SCORING..

a. Scoring will be done on the firing line as well as in the pits. Eachstudent will fire 38 rounds plus two sighter on the 300 yard line to check weapon's zero. Each round will be value at one point with a total value of 38 points. Passing score will be 80% of a "POSIBLE" SCORE, WHICH IS 28 HITS. Amiss will be scored as a zero. Final score will be determine by the pit score, and verifiers.

STALKING EXERCISES

1. The purpose of the stalking exercise is to give the sniper confidence in his ability to approach and occupy a firing postion without being observed.

2. DISCRIPTION:

Having studied a map (and aerial photograph,if available), individual students must stulk for a predesignated distance, which could be 1000 yards or more, depending on the area selected. All stalking exercises and tests should be approximately 1000 yards with a four hour time limit. The student must stail within 150 to 200 yards of two trained observers, who are scanning the area with binoculars, and fire two blanks without being detected.

The area used for a stalking exercise must be chosen with great care. An area in which a student must do the low crawl for the complete distance would be unsuitable. The following items should be considered:

1. As much of the area as possible should be visible to the observer. This forces the student to use the trrain properly, even when far from the observer's location.

2. Where possible, availble cover should decrease as the student nears the observer's position. This will enable the student to take chances early in the stalk and force him to move more carefully as he closes in on his firing postion.

3. The student must start the stalk in an area out of sight of the observer.

4. Boundaries must be established by means of natural features or the use of markers.

5. In a location near the jump off point for the stalk, the student is briefed on the following:

a. Aim of the exercise.

b. Boundaries.

c. Time limit (usualy 4 hours).

d. Standards to be achieved.

4. After the briefing, the students are dispatched at intervals to avoid congestion.

5. In addition to the two observers, there are two "walkers", equipped with radios, who will postion themselfs within the stalk area. If an observer sees a student, he will contact a walker by radio and direct him to within 5 feet of the students loction. Therefore, when a student is detected, the observer can immediately tell the student what gave him away.

6. When the student reaches his firing postion, which is within 150 to 200 yards of the observer, he will fire a blank round at the observer. This will tell the walker he is raedy to continue the rest of the exercise. The observer will then move to within 10 yards of the student. The observer will serch a 10 yard radius around the walker for the sniper student.

If the sniper is undetected the walker will tell the sniper to chamber another round and fire a second blank at the observer. If the sniper is still unseen, the walker will point to the sniper's postion, and the observer will serch for anything that indicates a human form, rifle, or equipment.

If the sniper still remains undetected, the walker will move in and put his hand on top of the student head. The obsever will again serch in detail. If the student is still not seen at this point, he must tell the walker which observer he shot and what he is doing. The observer waves his hat, scratches his face, or makes some kind of gesture that the student can identify when using his telescope.

The sniper student must then tell the walker the exact range, wind velocity, and windage applied to the scope.

If the sniper completes all of these steps correctly, he passes the stalk exercise.

A critique is conducted at the conclusion of the exercise, touching on main problems areas.

CREATING INTEREST:

To create interest and give the students practice in obsevering and stalking and stalking skills, one half of the class could be postioned to observe the conduct of the stalk. Seeing an error made is an effective way of teaching better stalking skills. When a student is caught, he should be sent to the observer post (OP) to observe the exercise.

1. All students to pass the stalking phase of the sniper course must acheved a total of the following:

a. Total of 63 stalking points.

b. Pass one stalk with a perfect score of 10 points.

2. SCORES. (If the students are caught prior to receving a score of 10 points)

1. 4 points for starting the stalk.

2. 5 points for reaching the F.F.P. and is detected by the observer (the student must be completely setup and ready to fire if not 4 points will be given to the sniper).

3. 7 points for reaching the F.F.P. undetected and getting one shot off and is detected by the observer.(if the student cannot discribe what the observer is doing 5 points will be awarded to the student.

4. 8 points for reaching the F.F.P. undetected, getting two shots off,and is detected by the observer. (if the student fails to discribe what the observer is doing 5 points will be awared to the sniper student.

5. 9 points for reaching the F.F.P. undetected, getting two shots off, the walker puts his hand on the student sniper's head and is detected by the observer(if the sniper cannot discribe what the observer is doing 5 points will be awarded to the student.

NOTE. One point will be taken from what ever score acheved for the following:

a. Incorrect windage.

b. Incorrect range setting.

c. Incorrect rifle setup(tape over muzzle, barrel of weapon laying on any kind of object, barrel is not free floating from not removing obstuction fron between the barrel and the stock of the weapon.

NOTE: A score of 0 will be awarded to any student who exceeds the boundaries of any stalk.

HIDE CONSTRUCTION EXERCISE

The purpose of the hide construction exercise is to show the sniper how to bui hide and remain undetected while being observed. The purpose of a hide is to camouflage a sniper or sniper team which is not in movement.

1. DESCRIPTION.

The sniper team is given 8 hours to build a temporary hide large enough to hold a sniper team with all their necesary equipment.

The hide area should be selected with great care. It can be in any type terrain, but there should be more than enough prospective spots in which to build a hide. The are should be easily bounded by left and right, far and near limits so that when the instructor points out the limits to the students, they can be easly and quickly identified. There should be enough tools(i.e.,axes,picks,shovels,and sandbags) availible to accommodate the entire class. There must be sufficient rations and water availible to the student to last the entire exercise, 9½ hours.

2. CONDUCT OF THE EXERCISE.

The students are issued a shovel, ax, pickax, and approximately 20 sandbags per team The students are brought to the area and briefed on the purpose of the exercise, their time limit for construction, and their area limits. The students are then allowed to begin construction of their hides.

NOTE: An instructor should be present at all time to act as an advisor.

At the end of 8 hours, the student's hides are all checked to ensure that they are complete. An Naval special warfare officer is brought out to act as an observer. He is placed in an area 300 yards from the hide area, where he starts his observation with binoculars and a 20x, spotting scope. The observer, after failing to find a hide, is brought forward 150 yards and again commences observation.

An instructor in the field (walker with radio) then moves to within 10 yards of a hide and informs the observer. The observer then tells the walker to have the sniper in the hide to load and fire a blank. If the sniper's muzzle blast is seen, or if the hide is seen due to improper construction, the team fails, but they remain in the hide. These procedures are repeated for all the sniper teams.

The observer is then brought down to within 25 yards of each hide to determine whether they can be seen with the naked eye at that distance.

The observer is not shown the hide. He must find it.

If the sniper team is located at 25 yards, it fails and is allowed to come out and see its discrepancies. If the team is not seen, it passes.

3. OTHER REQUIREMENTS.

The sniper team should also be required to fill out a range card and a sniper's log book and make a sketch. One way of helping them achieve this is to have an instructor shoeing"flash cards" from 150 yards away, begiing when the observer arrives and ending when the observer moves to within 25 yards. The sniper teams should record everything they see on the flash cards and anything going on at the observation post during the exercise.

4. STANDARDS.

The sniper teams are required to pass all phases in order to pass the exercise. All range cards , log books, and field sketches must be turned in for grading and a final determination of pass or fail.

RANGE ESTAMATION EXERCISES

Purpose of the range estimation exercises are to make the sniper proficient in accurately judging distance.

1. DESCRIPTION.

The student is taken to an observatio post, and different objects over distances of up to 1000 meters are indicated to him. After time for consideration, he writes down the estimated distance to each object. He may use only his binoculars and rifle telescope as aids, and he must estimate to within 10% of the correct range(a 6-foot man-size target should be utilized).

Each exercise must take place in a different area, offering a variety of terrain. The exercise areas should include dead ground as well as places where the student will be observing uphill or downhill. Extra objects should be selected in case those originally chosen cannot be seen due to weather or for other reasons.

2. CONDUCT OF THE EXERCISE.

1. The students are brought to the obervation post, issued a record card, and given a reveiw on the methods of judging distances and the causes of miscalculation. they are briefed on the following:

a. Aim of the exercise.

b. Reference points.

c. Time limit.

d. Standard to be achieved.

2. Students are spread out and the first object is indicated. The student will have 3 mimutes to estimate the distance and write it down. The sequence is repeated for a total of eight objects. The cards are collected, and the correct range to each object is given. The tnstructor points out in each case why the distance might be underestimated or overestimated. After correction, the cards are given back to the students after the instructor has recorded the scores of the students. In this way, the students retains a records of his performence.

3. STANDARDS. The student is deemed to have failed if he estimates three or more targets incorrectly

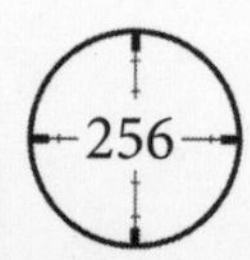

OBSEVATION EXERCISE

The purpose of the obsevation exercise is to practice the sniper's ability to observe an enemy and accurately record the results of his observations.

1. DISCRIPTION.

The student is given an arc of about 180 degrees to his front to observe for a period of not more than 40 minutes. He is issued a panoramic sketch or photograph of his arc and is expected to plot on the sketch or photo any objects he sees in his area.

Objects are so positioned as to be invisible to the naked eye, indistinguishable when using binoculars, but recognizable when using the spotting scope.

In choosingthe location for the exercise, the following points should be considered:

a. Number of objectsin the arc.(normaly 12 military items).

b. Time limit.

c. Equipment which they are allowed to used (binos, spotting scope).

d. Standard to be attained.

Each student takes up the prone postion on the observatio line and is issued a sketch or photo of the area.

The instructor staff is availible to answer any questions about the photo or sketch if a student is confused.

If the class is large, the observation line could be broken into a right side and a left side. A student could spend the first 20 minutes in one half and then move to the other. This ensures that he sees all the ground in the arc.

At the end of 40 minutes, all sheets are collected and the students are shown the loction of each object. This is best done by the student staying on their postions and watching while the instructor points to each object. In this way, the student will see why he over looked the object, even though it was visible.

A critique is then held, bring out the main points.

2. SCORING.

Students are given half a point for each object correctly plotted and another point for naming the object correctly.

3. STANDARDS.

The students is deemed to have failed if he scores less than 8 points out of 12 points.

MEMORY EXERCISE (KIM'S GAME)

The purpose of the memory exercise is to teach the sniper student to observe and remember a number of unrelated objects. In combat, the sniper brequires a good memory in order to report facts accurately, because he may not be in a postion to write them down. The Kim's game is to help the student in observation techniques. The better he does on the Kim's games, the more confident he will be during the observation exercises.

1. PREPARATION.

a. The instructor places 12 small objects on a table (usualy 12 military objects) They could be anything from a paper clip to a 40mm round. He notes the name of each object and it's color and it's most distingushable features (color, shape, size, lettering, etc).

b. The student are placed in a circle around a covered table and told the purpose of the exercise.

2. CONDUCT OF THE EXERCISE.

a. The instructor tells the students there are 12 objects on the table. He explains that they have a small amount of time to look and a slightly longer amount of time to write. This could range from 2 minutes to look and 2½ minutes to write on the first exercise to 20 seconds to look and 30 seconds to write on the last exercise.

b. After the "looking" time limit is up, the students are given a time limit to write down what they saw.

c. Papers are cllected, and the objects are again displayed to show the students what they missed.

3. DEGREE OF DIFFICULTY.

a. Successive games can be increased in difficulty by:

a. Shorting the limits to look and write.

b. Creating distractions, such as music, noise, etc.

c. Sending the students on a short run after they view the objects, then given them a shorter amount of time to write.

d. Having the students go on a scheduled field craft exercise after viewing the objects, then after returning (1 to 2 hours later), having them write down what they saw in the Kim's game.

CAMOUFLAGE AND CONCEALMENT EXERCISES

Camouflage and concealment exercises are held to help the sniper student to select final firing postions.

1. DESCRIPTION.

The student conceals himself within 200 yards of an observer, who, using binoculars, tries to find the student. The student must be able to fire blank ammuntion at the observer without being seen, and have the correct elevation and windage on his sights. The student must remain unseen throughout the conduct of the exercise.

In choosing the location for the exercise, the instructor ensures that certain condistions are met. These are:

a. There must be adequate space to ensure students are not crowed together in the area. There should be at lease twice the number of potential postions as there are students. Once the area has been established the limits should be marked in some manner (e.g.,flags,trees,prominent features,etc). Students should then be allowed to choose their final firing postion.

b. The observer must be located where he can see the entire problem area.

As therewill be serveral concealment exercise throughout the sniper course, different types of terrain should be choosen in order that the student may practice concealment in varied condistions. For instance, one exercise could place in fairly open area, one along a wood line, one in shrubs, and another in hilly or rough terrain.

2. CONDUCT OF THE EXERCISE.

The sniper is given a specified area with boundaries in which to conceal himself properly. The observers turn their backs to the area and allow the students 5 minutes to conceal themselves. Athte end of 5 minutes, the obersvers turn and commence observation in their search for concealed snipers. This observation should last approximately one-half hour.

At the conclusion of the observation, the observer will instruct, by radio, one of the two walkers in the field to move to within the 10 meters of one of the snipers. The sniper is given one blank. If the sniper cannot be seen by the observer after moving to within 10 meters, the walker will tell him to load and fire his blank.

The observer is looking for muzzle blast, vegitation flying after the shot, and movement by the sniper before and after he fires.

If the student cannot be seen, the walker then extends his arm in the sniper's direction, indicating his postion. If the sniper still remains unseen after indiction, the walker goes to the sniper's postion and places his hand, palm facing the observer, directly on top of the sniper's head.

If the sniper passes all the above, he must then state his elevation, windage, and what type of movement the observer is making.

3. STANDARDS.

The sniper student must meet all the above condtions to receive a passing score.

The student is deemed to have failed if he does not recieve a passing score of 5 points. A total of 10 points must be accumalated in order to pass the camouflage and concealment exercises. (8 exercises will be given throughout the course).

مرحبا بكم في العراق
Welcome To Iraq

SCOUT/SNIPER OBSERVER HALO/HAHO INSERTION EQUIPMENT

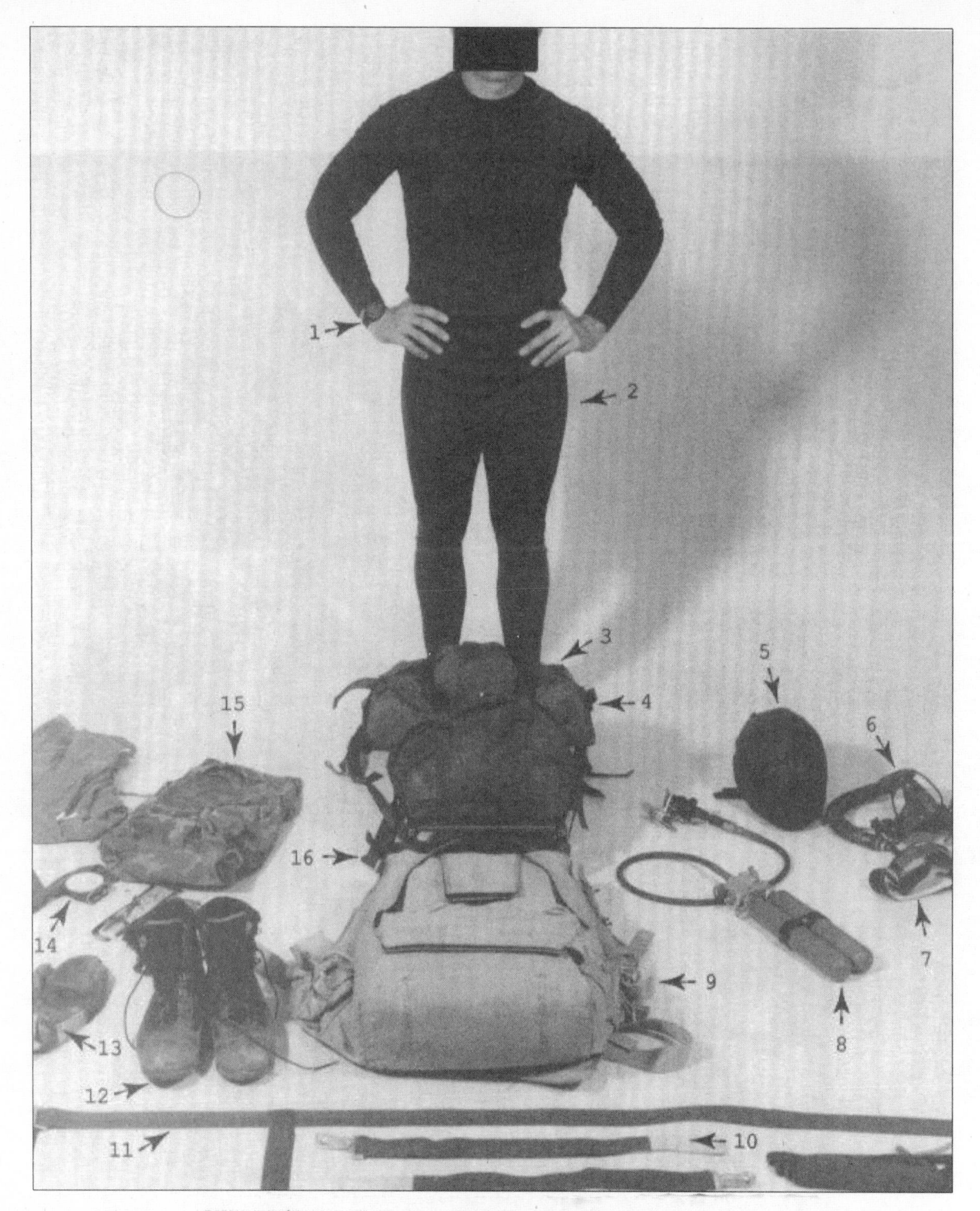

SNIPER/OBSERVER EQUIPMENT, READY TO BE DONNED

1. WATCH
2. THERMAL UNDERWEAR FOR HIGH-ALTITUDE JUMPS
3. SECOND-LINE EQUIPMENT BAG
4. RUCKSACK
5. HELMET
6. OXYGEN MASK
7. GOGGLES
8. OXYGEN BOTTLE
9. MTIX PARACHUTE
10. QUICK RELEASES
11. "H" HARNESS
12. BOOTS
13. HIGH-ALTITUDE-JUMP GLOVES (MITTENS)
14. ALTIMETER
15. FIELD UNIFORM
16. HOOK KNIFE

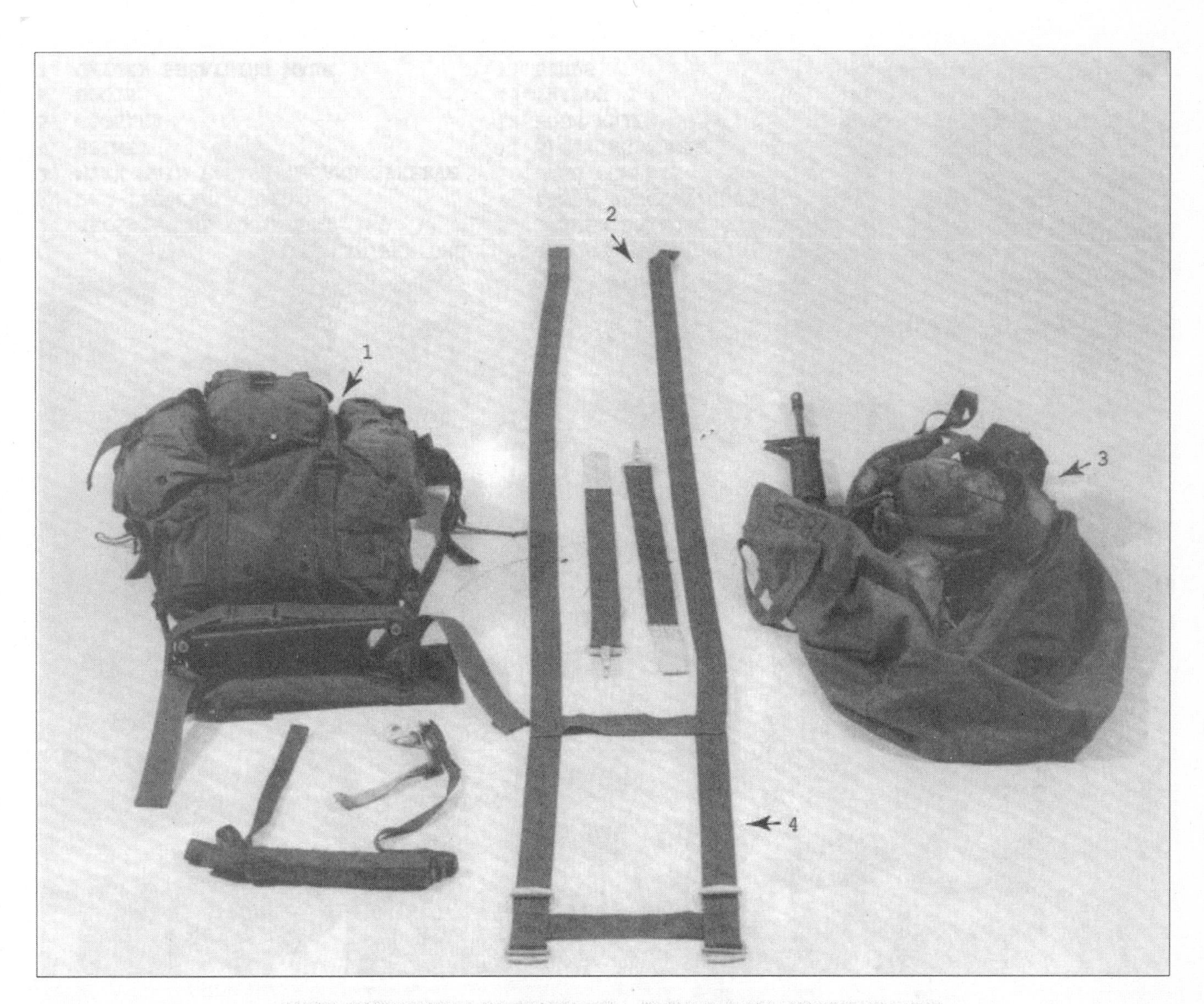

SNIPER/OBSERVER RUCKSACK AND SECOND-LINE EQUIPMENT BAG

1. RUCKSACK
2. QUICK RELEASES
3. SECOND-LINE EQUIPMENT
4. "H" HARNESS

SNIPER/OBSERVER MISSION, ESSENTIAL EQUIPMENT

1. SECOND-LINE EQUIPMENT
2. PRC-113 FIELD RADIO
3. MTIX MAIN PARACHUTE AND RESERVE
4. HELMET
5. GOGGLES
6. BOOTS
7. OXYGEN BREATHING MASK
8. OXYGEN BAILOUT BOTTLE
9. LASER RANGE FINDER
10. HAHO COMPASS
11. SPOTTING SCOPE
12. HOOK KNIFE
13. TRIPON
14. BINDS
15. "H" HARNESS
16. 30-ROUND MAGAZINES
17. PRIMARY WEAPON
18. CAMMIE PAINT
19. BUG REPELLANT
20. OBSERVER LOGBOOK
21. VDT LIFE JACKET
22. OBSERVER FIELD UNIFORM
23. SECONDARY WEAPON

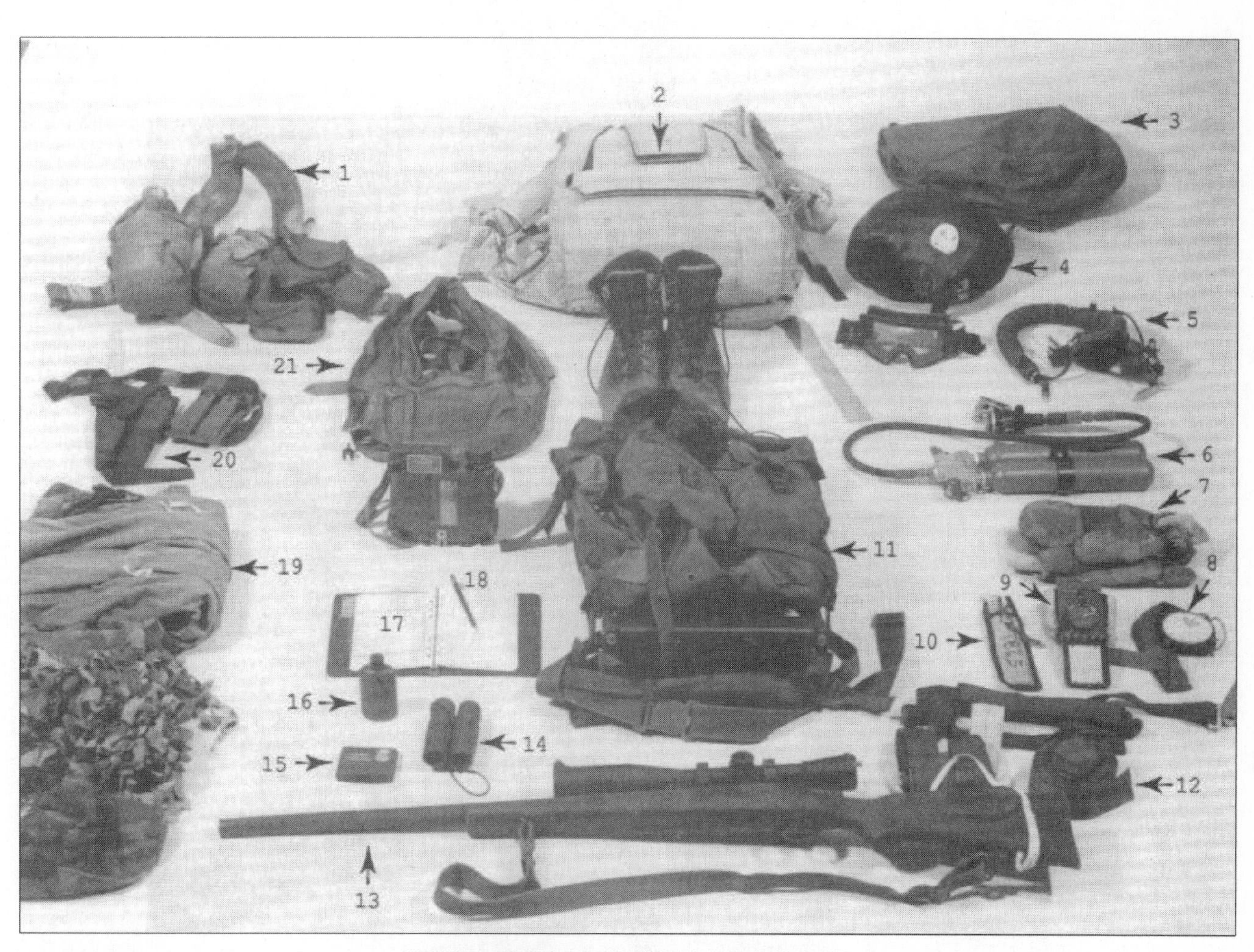

SNIPER MISSION INSERTION EQUIPMENT

1. SECOND-LINE EQUIPMENT
2. MTIX MAIN PARACHUTE, AND RESERVE
3. SECOND LINE EQUIPMENT BAG
4. HELMET
5. OXYGEN BREATHING MASK
6. OXYGEN BAILOUT BOTTLE
7. HAHO GOLVES
8. ALTIMETER
9. HAHO COMPASS
10. HOOK KNIFE
11. RUCKSACK
12. "H" HARNESS AND QUICK RELEASES
13. PRIMARY WEAPON
14. BIND
15. CAMMIE PAINT
16. BUG REPELLANT
17. SNIPER LOGBOOK
18. PEN
19. SNIPER FIELD UNIFORM
20. SECONDARY WEAPON
21. VDT LIFE JACKET

SNIPER EQUIPMENT FULLY DONNED, FRONT VIEW

SNIPER EQUIPMENT FULLY DONNED, SIDE VIEW

SNIPER OBSERVER WITH HAHO EQUIPMENT DONNED, FRONT VIEW

PROPER WEAPONS STORAGE, SIDE VIEW

EQUIPMENT TO BE PREPPED FOR WATERBORNE INSERTION. NOTE THAT THE SCOPE HAS BEEN REMOVED FROM THE RIFLE AND WATERPROOFED; TORQUE WRENCH REQUIRED TO REMOUNT IN ORDER TO MAINTAIN ZERO.

SCOUT/SNIPER WATERBORNE INSERTION EQUIPMENT

FULLY SUITED UP, LESS RIFLE (FRONT VIEW)

FULLY SUITED UP, LESS RIFLE (REAR VIEW)

READY-TO-DON DIVING GEAR (FRONT VIEW, WETSUIT UNDER FLIGHT SUIT)

READY-TO-DON DIVING GEAR (REAR VIEW, WETSUIT UNDER FLIGHT SUIT)

SEAL (KNEELING) WITHOUT GHILLIE SUIT AT LEFT; SEAL (KNEELING) IN GHILLIE SUIT AT RIGHT

MODIFIED GHILLIE SUIT, REAR VIEW

GHILLIE SUIT, REAR VIEW

VARIOUS SEAL SNIPER WEAPONS AND EQUIPMENT

SEAL WITHOUT GHILLIE SUIT

SCOUT/SNIPER FIELD GEAR, READY FOR DONNING

M-16 WITH NIGHT-VISION SCOPE

SEAL SNIPER WITH GHILLIE SUIT

SCOUT/SNIPER, FULLY SUITED UP

SNIPER IN GHILLIE SUIT BLENDS IN WITH TREE

SCOUT/SNIPER OBSERVER FIELD GEAR, READY FOR DONNING

M-16 WITH NIGHT-VISION SCOPE

SEAL SNIPER WITH GHILLIE SUIT

SEAL (STANDING) WITHOUT SNIPER GHILLIE SUIT ON LEFT; SNIPER (STANDING) IN GHILLIE SUIT AT RIGHT

GHILLIE SUIT

VARIOUS SEAL SNIPER WEAPONS AND EQUIPMENT

Enclosure 1 6 Week Training schedule.

DAY 1.

CLASS ROOM..................................(AM)
CLASS ROOM..................................(PM)

DAY 2.

KNOWN DISTANCE RANGE........................(AM)
CLASS ROOM..................................(PM)
KIM'S GAMES # 1.............................(PM)
CLASS ROOM..................................(PM)

DAY 3.

KNOWN DISTANCE RANGE........................(AM)
RANGE ESTIMATION EXERCISE # 1...............(PM)
KIM'S GAMES # 2.............................(PM)
CLASS ROOM..................................(PM)

DAY 4.

KNOWN DISTANCE RANGE........................(AM)
OBSERVATION EXERCISE # 2....................(PM)
RANGE ESTIMATION EXERCISE # 2...............(PM)
KIM'S GAMES # 3.............................(PM)
CLASS ROOM..................................(PM)

Day 5.

KNOWN DISTANCE RANGE........................(AM)
OBSERVATION EXERCISE # 3....................(PM)
RANGE ESTIMATION EXERCISE # 3...............(PM)
KIM'S GAMES # 4.............................(PM)
CLASS ROOM..................................(PM)

DAY 6.

LAND NAVIGATION EXERCISE TEST...............(AM)
KNOWN DISTANCE RANGE,
INTRODUCTION TO THE 50 CAL SWS..............(PM)
CAMMIE AND CONCEALMENT EXERCISE # 1.........(PM)

DAY 7.

OFF

DAY 8.

STALK # 1.....................................(AM)
KNOWN DISTANCE RANGE..........................(PM)
OBSERVATION EXERCISE # 4......................(PM)
RANGE ESTIMATION EXERCISE # 4.................(PM)
CLASS ROOM....................................(PM)

DAY 9.

KNOWN DISTANCE RANGE..........................(AM)
RANGE CARDS AND
FIELD SKETCHING EXERCISE # 1..................(PM)
CLASS ROOM....................................(PM)

DAY 10.

KNOWN DISTANCE RANGE..........................(AM)
OBSERVATION EXERCISE # 5......................(PM)
RANGE ESTIMATION EXERCISE # 5.................(PM)
KIM'S GAMES # 5...............................(PM)
CLASS ROOM....................................(PM)

DAY 11.

KNOWN DISTANCE RANGE..........................(AM)
OBSERVATION EXERCISE # 6......................(PM)
RANGE ESTIMATION EXERCISE # 6.................(PM)
KIM'S GAMES # 6...............................(PM)
CLASS ROOM....................................(PM)

DAY 12.

STALK # 2.....................................(AM)
KNOWN DISTANCE RANGE..........................(PM)

DAY 13.

LAND NAVIGATION EXERCISE # 2..................(AM)
50 CAL SWS, UNKNOWN DISTANCE RANGE............(PM)

DAY 14.

OFF

DAY 15.

STALK # 3....................................(AM)
UNKNOWN DISTANCE RANGE.......................(PM)
RANGE ESTIMATION EXERISE # 7.................(PM)
OBSERVATION EXERCISE # 7.....................(PM)
CLASS ROOM...................................(PM)

DAY 16.

UNKNOWN DISTANCE RANGE.......................(AM)
KIM'S GAME # 7...............................(PM)
CLASS ROOM (LZ/PZ,CALL FOR FIRE).............(PM)

DAY 17.

UNKNOWN DISTANCE RANGE.......................(AM)
HELO INSERTIONS AND
CALL FOR FIRE................................(PM)

DAY 18.

STALK # 4....................................(AM)
UNKNOWN DISTANCE RANGE.......................(PM)
NIGHT SHOOT, KNOWN DISTANCE RANGE.............(PM)

DAY 19.

STALK # 5....................................(AM)
UNKNOWN DISTANCE RANGE.......................(PM)

DAY 20.

LAND NAVIGATION EXERCISE # 3.................(AM)
50 CAL SWS, UNKNOWN DISTANCE RANGE...........(PM)

DAY 21.

OFF

DAY 22.

STALK # 6..(AM)
UNKNOWN DISTANCE RANGE...........................(PM)
KIM'S GAME # 8...................................(PM)
CLASS ROOM.......................................(PM)

DAY 23.

KNOWN DISTANCE RANGE.............................(AM)
OBSERVATION EXERCISE # 8.........................(PM)
RANGE ESTIMATION EXERCISE # 8....................(PM)
FIELD SKETCHING EXERCISE # 2.....................(PM)
CLASS ROOM.......................................(PM)

DAY 24.

KNOWN DISTANCE RANGE.............................(AM)
KIM'S GAMES # 9..................................(PM)
NIGHT SHOOT, UNKNOWN DISTANCE RANGE..............(PM)

DAY 25.

UNKNOWN DISTANCE RANGE...........................(AM)
FIELD SKETCHING AND
RANGE CARDS EXERCISE # 3.........................(PM)

DAY 26.

STALK # 7..(AM)
KNOWN DISTANCE RANGE.............................(PM)

DAY 27.

LAND NAVIGATION EXERCISE # 4.....................(AM)
50 CAL SWS, UNKNOWN DISTANCE RANGE...............(PM)

DAY 28.

OFF

DAY 29.

STALK # 8......................................(AM)
KNOWN DISTANCE RANGE...........................(PM)
CLASS ROOM.....................................(PM)

DAY 30.

UNKNOWN DISTANCE RANGE.........................(AM)
RANGE ESTIMATION EXERCISE # 9..................(PM)

DAY 31.

UNKNOWN DISTANCE RANGE.........................(AM)
OBSERVATION EXERCISE # 9.......................(PM)
KIM'S GAMES # 10...............................(PM)
CLASS ROOM.....................................(PM)

DAY 32.

KNOWN DISTANCE RANGE...........................(AM)
RANGE ESTIMATION EXERCISE # 10.................(PM)
OBSERVATION EXERCISE # 10......................(PM)
CLASS ROOM

DAY 33.

STALK # 10.....................................(AM)
INTERNAL SECURITY SHOOT,
UNKNOWN DISTANCE RANGE.........................(PM)
KNIGHT SHOOT,KNOWN DISTANCE RANGE..............(PM)

DAY 34.

LAND NAVIGATION EXERCISE # 5...................(AM)
HELO INSERTION/EXTRACTIONS.....................(PM)

DAY 35.

OFF

DAY 36.

STALK # 11..(AM)
KNOWN DISTANCE RANGE..............................(PM)
OBSERVATION EXERCISE # 11.........................(PM)
RANGE ESTAMATION EXERCISE # 11....................(PM)

DAY 37.

PRACTICE MARKSMANSHIP TEST
KNOWN DISTANCE RANGE..............................(AM)
PRACTICE MARKSMANSHIP TEST
UNKNOWN DISTANCE..................................(PM)
WRITTEN TEST......................................(PM)

DAY 38.

MARKSMANSHIP TEST
KNOWN DISTANCE RANGE..............................(AM)
MARKSMANSHIP TEST
UNKNOWN DISTANCE RANGE............................(PM)
MISSION TASKING (FTX).............................(PM)

DAY 39.

WARNING ORDER.....................................(AM)
PATROL ORDER......................................(PM)
FTX (HELO INSERTION)..............................(PM)

DAY 40.

FTX...(AM/PM)

DAY 41.

FTX (HELO EXTRACTION).............................(AM)
FTX (HELO INSERTION)..............................(PM)

DAY 42.

FTX (HELO EXTRACTION).............................(AM)
CAMP CLEAN UP.....................................(AM)
EQUIPMENT CLEAN UP AND PREP.......................(PM)
TRANSIT ....(PM)

Enclosure 2 Required Equipment Load Out.

CHULK BOARD....................................1ea
CHULK (VARIOUS COLORS).........................1ea
CHULK ERASER...................................1ea
PAPER NOTE BOOK PADS..........................10ea
PENS...3ea (boxes)
PENCILS..3ea (boxes)
GREASE PENCELS.................................3ea (boxes)
SLIDE PROJECTOR................................1ea
EXTENSION CORD.................................1ea
200 YARD PAPER TARGETS.......................100ea
200 YARD REPAIR FACES........................100ea
BLACK TARGET PASTIES..........................10ea (rolls)
WHITE TARGET PASTIES..........................10ea (rolls)
TARGET DISKS 1', 3', 6'........................2ea (boxes)
TARGET PASTE...................................6ea (boxes)
TARGET FBI.....................................4ea (boxes)
TARGET DOG.....................................2ea (boxes)
TOMATO STAKES 6 FT............................50ea
PAINT SPRAY WHITE..............................4ea
PAINT SPRAY BLACK..............................4ea
STAPLE GUN.....................................2ea
STAPLES..3ea (boxes)
HAMMER...2ea
NAILS..300ea
RANGE BOX......................................1ea
BREAK FREE 1 QT BOTTLE.........................2ea
GUN CLEANING EQUIPMENT BOX.....................2ea
TOOLS VARIOUS..................................1ea (boxes)
RAGS...1ea (bundel)
TORQUE WRENCH..................................1ea
KIMS GAMES MATERIAL...........................12 (games)
OBSERVATION EXERCISE MATERIAL.................12 (exercises)
CHEM LIGHTS..................................100ea
WATER 5 GAL CONTAINERS.........................2ea
BATTERIES AA...................................2ea (boxes)
BATTERIES BA5590 (NVG)........................20ea
MX-300 HAND RADIO..............................6ea
MX-300 BATTERY CHARGER.........................1ea
PRC-113..2ea (complete)
PRC-117..5ea (complete)
HAND SMOKE GERNADE YELLOW......................10ea
HAND SMOKE GERNADE GREEN.......................10ea
HAND SMOKE GERNADE RED.........................10ea
HAND FLARE POP-UP RED STAR CLUSTER.............1ea (case)
HAND FLARE POP-UP GREEN STAR CLUSTER...........1ea (case)
HAND FLARE POP-UP WHITE STAR CLUSTER...........1ea (case)

HAND FLARE POP-UP WHITE PARACHUTE.........1ea (case)
BOOBIE TRAP FLARE POP-UP.................6ea (boxes)
AMMO BOXES (land navigation points).......12ea
MAP 1 : 25000.............................15ea (operation area)
MAP 1 : 50000.............................15ea (operation area)
PROTRACTOR 1 : 25000/1 : 50000............15ea
OVERLAY PAPER.............................1ea (box)
MAGIC MARKERS VARIOUS COLORS..............2ea
ACETATE PAPER.............................1ea (roll)
BINOS.....................................4ea
SPOTTING SCOPES...........................4ea
NIGHT VISION GOGGLE.......................4ea
NIGHT VISION SCOPE PVS-4/M-845............2ea
M-86 SNIPER RIFLE.........................4ea
M-14 RIFLE................................4ea
BERRETTA 92FB.............................2ea
EAR PROTECTION............................2ea (boxes)
PARA CORD 550.............................1ea (roll)
NYLON TUBULAR 1'..........................1ea (roll)
RUBBER BANDS..............................2ea (boxes)
LANTERN COLMAN............................2ea
GENERATOR PORTABLE........................1ea
GAS CAN 5 GAL.............................1ea
MOTOR OIL 1 QT............................2ea
TRASH BAGS LARGE..........................3ea (boxes)
RIGGER'S TAPE.............................4ea (rolls)
TENT 30 MAN...............................1ea
I.R. NETTING WOOLAND/DESERT...............1ea (terrain dictates)
CHAIR PORTABLE............................4ea
TABLE PORTABLE............................1ea
AMMUNITION 7.62mm M-118...................as required
AMMUNITION 7.62mm M-80....................as required
AMMUNITION 50 CAL.........................as required
TRANSPORTATION............................as required
GOVERNMENT CREDIT CARD....................1ea
ENTRENCHING TOOL..........................4ea
AX..1ea
SLEDGE HAMMER.............................1ea
300/600 YARD PAPER TARGETS................50ea
300/600 YARD PAPER REPAIR CENTERS.........100ea
DEMO BOX..................................1ea
ELECTRIC BLASTING CAPS....................100ea
ARTILLERY SIMULATORS......................2ea (cases)
GERNADE HAND SIMULATORS...................2ea (cases)
GERNADE HAND SIMULATORS C.S...............1ea (case)
BUNGIE CORD...............................1ea (roll)
PLY WOOD..................................2ea (standard sheet)
WOOD 2X4..................................4ea
PAPER ROLL (white) 6" X 25"...............1ea
WATER PROOF BAG (american safety bag).....2ea

Enclosure 3 Required Student Load Out.

CAMMIES 2ea
CAMMIE HAT..................................1ea
JUNGLE HAT..................................2ea
BOOTS.......................................2ea
FLIGHT GLOVES...............................1ea
WARM CLOTHS.................................(as required)
CAMMIE PAINT................................(as required)
STANDARD WEB GEAR (M-14)....................1ea
SNAP LINK...................................3ea
COMPASS.....................................1ea
FLASH LIGHT.................................1ea
FAST ROPE GLOVES............................1ea
RUCK SACK (small)...........................1ea
HOLSTER (9mm)...............................1ea
MAGS (9mm)..................................3ea
MAGS (M-14).................................8ea
M-86 SNIPER RIFLE...........................1ea
M-14 RIFLE..................................1ea
RIFLE CLEANING GEAR (M-14/M-86).............1ea
BINOS.......................................1ea
SPOTTING SCOPE..............................1ea
PARA BAG....................................1ea
NOTE BOOK...................................1ea
SHOOTER'S DATA BOOK.........................2ea
PENCIL/PEN..................................(as required)
PAPER.......................................(as required)
GILLIE SUIT MAKING MATERIAL.................(as required)
STROBE LIGHT W/ I.R. COVER..................1ea
MK-13 FLARE.................................2ea
CHEM LIGHT..................................6ea
PONCHO......................................1ea
RAIN GEAR...................................1ea
PONCHO LINER................................1ea
SLEEPING BAG................................(as required)
UDT LIFE JACKET.............................1ea
WATER PROOF BAG (AMERICAN SAFETY BAG)....1ea
WET SUIT....................................(as required)
BUG SPRAY...................................(as required)
KNIFE.......................................1ea
TOWEL.......................................1ea
NIGHT VISION SCOPE (PVS-4/M-845
W/ M-14 MOUNT...............................1ea
PROTECTIVE CARRYING CASE M-14/M-86.........1ea
ENTRENCHING TOOL W/ COVER...................1ea
DRAG BAG....................................1ea
BIPOD.......................................1ea

HAND FLARE POP-UP WHITE PARACHUTE.........1ea (case)
BOOBIE TRAP FLARE POP-UP..................6ea (boxes)
AMMO BOXES (land navigation points).......12ea
MAP 1 : 25000.............................15ea (operation area)
MAP 1 : 50000.............................15ea (operation area)
PROTRACTOR 1 : 25000/1 : 50000............15ea
OVERLAY PAPER.............................1ea (box)
MAGIC MARKERS VARIOUS COLORS..............2ea
ACETATE PAPER.............................1ea (roll)
BINOS.....................................4ea
SPOTTING SCOPES...........................4ea
NIGHT VISION GOGGLE.......................4ea
NIGHT VISION SCOPE PVS-4/M-845............2ea
M-86 SNIPER RIFLE.........................4ea
M-14 RIFLE................................4ea
BERRETTA 92FB.............................2ea
EAR PROTECTION............................2ea (boxes)
PARA CORD 550.............................1ea (roll)
NYLON TUBULAR 1'..........................1ea (roll)
RUBBER BANDS..............................2ea (boxes)
LANTERN COLMAN............................2ea
GENERATOR PORTABLE........................1ea
GAS CAN 5 GAL.............................1ea
MOTOR OIL 1 QT............................2ea
TRASH BAGS LARGE..........................3ea (boxes)
RIGGER'S TAPE.............................4ea (rolls)
TENT 30 MAN...............................1ea
I.R. NETTING WOOLAND/DESERT...............1ea (terrain dictates)
CHAIR PORTABLE............................4ea
TABLE PORTABLE............................1ea
AMMUNITION 7.62mm M-118...................as required
AMMUNITION 7.62mm M-80....................as required
AMMUNITION 50 CAL.........................as required
TRANSPORTATION............................as required
GOVERNMENT CREDIT CARD....................1ea
ENTRENCHING TOOL..........................4ea
AX..1ea
SLEDGE HAMMER.............................1ea
300/600 YARD PAPER TARGETS................50ea
300/600 YARD PAPER REPAIR CENTERS.........100ea
DEMO BOX..................................1ea
ELECTRIC BLASTING CAPS....................100ea
ARTILLERY SIMULATORS......................2ea (cases)
GERNADE HAND SIMULATORS...................2ea (cases)
GERNADE HAND SIMULATORS C.S...............1ea (case)
BUNGIE CORD...............................1ea (roll)
PLY WOOD..................................2ea (standard sheet)
WOOD 2X4..................................4ea
PAPER ROLL (white) 6" X 25"...............1ea
WATER PROOF BAG (american safety bag).....2ea

Enclosure 4 Marksmanship Test.

1. The purpose of the marksmanship test is to evaluate the student's ability to engage targets at various ranges, scoring one point per hit with 80% accuracy.

2. The student will be required to engage stationary targets at ranges from 300 to 1000 yards, moving targets 300 t0 600 yards and pop-up targets from 300 to 800 yards and must get at lease 80% of the total rounds fired.

3. Needed equipment.

a. Communications equipment.....................MX-300R (4ea)

b. Score cards for the pits and the line (Only the pit score is valid.) Verifiers should be present (2ea).

c. Range Safety Officer.

d. Corpsman.

e. Emergency vehicle.

f. 1000 yard known distance range.

g. Range Guards (if required)

3. CONDUCT OF ENGAGING STATIONARY TARGETS FROM 300 TO 1000 YARDS.

a. Each team will be assigned a block of eight targets, each block of which will be designated with the left and right limits marked with a 6-foot target mounted in two respective carriages. Thus, the right limit for one block will also serve as the left limit of the next block. The following targets will serve as left and right limits respectively: 1, 8, 15, 22, 29, 36, and 43. The stationary target will be mounted in the left limit target carriage of each block.

b. The first stage of fire at each yard line (300, 500, 600, 700, 800, 900 and 1000) will be stationary targets from the supported prone position. Commands will be given from the center of the line by the range safety officer to load one round. The sniper and partner will have three mimutes to judge wind, light conditions, proper elevation hold, and fire three rounds with the target being pulled and marked after each shot.

(2) Target size formula - yard line x 2 = target diameter.
(Example 400 yard line - 4 x 2 = 8 inches in target diameter).

The shooter will engage two pop-up target on each yard line while his partner calls wind and recored all information in his data book.

c. Pop-up targets will not be engaged past 800 yards. Therefore, five rounds will be fired and scored on stationary targets at 900 and 1000 yards.

4. <u>TEST SCORING.SCORING</u>.

a. Scouring will be done on the firing line and in the pits. Each student will fire 44 rounds plus two sighter on the 300 yard line to check weapon's zero. Each round will be valued at one point with a total value of 44 points. Passing score will be 80% of a "POSSIBLE" score, or 38 hits. A miss will be scored as a zero. Final score will be determine by the pit score, and verifiers.

Enclosure 5 Observation Test.

1. The purpose of the observation exercise is to sharpen the sniper's ability to observe an enemy and accurately record the results of his observations.

2. CONDUCT OF THE OBSERVATION EXERCISE.

a. The student is given an arc of about 180 degrees to his front to observe for a period of not more than 40 minutes. He is issued a panoramic sketch or photograph of his arc and is expected to plot on the sketch or photo any objects he sees in his area.

b. Objects are so positioned as to be invisible to the naked eye, indistinguishable when using binoculars, but recognizable when using the spotting scope.

c. In choosing the location for the exercise, the following points should be considered:

(1) Number of objects in the arc. (normally 12 military items).

(2) Time limit. (40 minutes).

(3) Equipment which they are allowed to use. (binos, spotting scope).

(4) Standard to be attained. (80%)

d. Each student takes up the prone postion on the observation line and is issued a sketch or photo of the area. The instructor staff is availible to answer any questions about the photo or sketch if a student is confused.

If the class is large, the observation line could be broken into a right side and a left side. A student could spend the first 20 minutes in one half and then move to the other. This ensures that he sees all the ground in the arc.

At the end of 40 minutes, all sheets are collected and the students are shown the location of each object. This is best done by the students staying on their positions and watching while the instructor points to each object. In this way, the student will see why he overlooked the object, even though it was visible.

e. A critique is then held, bringing out the main points.

3. SCORING.

a. Student are given half a point for each object correctly plotted and another point for naming the object correctly.

4. STANDARDS.

a. The student is deemed to have failed if he scores less than 8 points out of 12 points.

b. During the duration of the 6 weeks of training 11 excercises will be conducted. The student must pass with 80% accuracy 8 of the 11 observation exercises in order to successfully pass the observation test.

Enclosure 6 Range Estimation.

1. Purpose of the range estimation exercises is to make the sniper proficient in accurately judging distance.

1. CONDUCT OF RANGE ESTIMATION EXERCISE.

a. The student is taken to an observation post, and different objects over distances of up to 1000 meters are indicated to him. After time for consideration, the student writes down the estimated distance to each object. He may use only his binoculars and rifle telescope as aids, and he must estimate to within 10% of the correct range (a 6-foot man-size target should be utilized).

b. Each exercise (11 ea.) must take place in a different area, offering a variety of terrain. The exercise areas should include dead ground as well as places where the student will be observing uphill or downhill. Extra objects should be selected in case those originally chosen cannot be seen due to weather or for other reasons.

b. The students are brought to the obervation post, issued a record card, and given a review on the methods of judging distances and the causes of miscalculation. They are then briefed on the following:

(1) Aim of the exercise.

(2) Reference points.

(3) Time limit. (3 minutes per object)

(4) Standard to be achieved. (80%)

c. Students are spread out and the first object is indicated. The student will have 3 minutes to estimate the distance and write it down. The sequence is repeated for a total of eight objects. The cards are collected, and the correct range to each object is given. The instructor points out in each case why the distance might be underestimated or overestimated. After correction, the cards are given back to the students after the instructor has recorded the scores of the students. In this way, the students retains a records of his performance.

d. SCORING.

a. The student is deemed to have failed if he estimates three or more targets incorrectly.

b. During the duration of the 6 weeks of training 11 excercises will be conducted, the student must pass with 80% accuracy 8 of the 11 range estimation excercises in order to successfully pass the range estimation test.

Enclosure 7 Stalking Test.

1. The purpose of the stalking exercise is to give the sniper confidence in his ability to approach and occupy a firing postion without being observed.

2. CONDUCT OF STALKING EXERCISE.

a. Having studied a map (and aerial photograph, if available), individual students must stalk for a predesignated distance, which could be 1000 yards or more, depending on the area selected. All stalking exercises should be approximately 1000 yards with a four hour time limit. The student must stalk to within 200 yards (+ or - 10%) of two trained observers (who are scanning the area with binocuiars) and fire two blanks without being detected.

b. The area used for a stalking exercise must be chosen with great care. An area in which a student must do the low crawl for the complete distance would be unsuitable. The following items should be considered:

(1) As much of the area as possible should be visible to the observer. This forces the student to use the terrain properly, even when far from the observer's location.

(2) Where possible, available cover should decrease as the student nears the observer's position. This will enable the student to take chances early in the stalk and force him to move more carefully as he closes in on his firing postion.

(3) The student must start the stalk in an area out of sight of the observer.

(4) Boundaries must be established by means of natural features or the use of markers.

(5) In a location near the jump off point for the stalk, the student is briefed on the following:

a. Aim of the exercise.

b. Boundaries.

c. Time limit (usually 4 hours).

d. Standards to be achieved. (63 points)

(6) After the briefing, the students are dispatched at intervals to avoid congestion.

(7) In addition to the two observers, there are two "walkers" equipped with radios, who will postion themselves within the stalk area. If an observer sees a student, he will contact a walker by radio and direct him to within 5 feet of the student's location. Therefore, when a student is detected, the observer can immediately tell the student what gave him away.

(8) When the student reaches his firing position, which is within 200 yards of the observer, he will fire a blank round at the observer. This will tell the walker he is ready to continue the rest of the exercise. The observer will then move to within 10 yards of the student. The observer will search a 10 yard radius around the walker for the sniper student.

(9) If the sniper is undetected, the walker will tell the sniper to chamber another round and fire a second blank at the observer. If the sniper is still unseen, the walker will point to the sniper's position, and the observer will search for anything that indicates a human form, rifle, or equipment.

(10) If the sniper still remains undetected, the walker will move in and put his hand on top of the student's head. The observer will again search in detail. If the student is still not seen at this point, he must tell the walker which observer he shot and what he is doing. The observer waves his hat, scratches his face, or makes some kind of gesture that the student can identify when using his telescope.

(11) The sniper student must then tell the walker the exact range, wind velocity, and windage applied to the scope. If the sniper completes all of these steps correctly he passes the stalk exercise.

(12) A critique is conducted at the conclusion of the exercise, touching on main problem areas.

3. CREATING INTEREST.

To create interest and give the students practice in observing and stalking and stalking skills, one half of the class could be positioned to observe the conduct of the stalk. Seeing an error made is an effective way of teaching better stalking skills. When a student is caught, he should be sent to the observer post (OP) to observe the exercise.

ALSO AVAILABLE

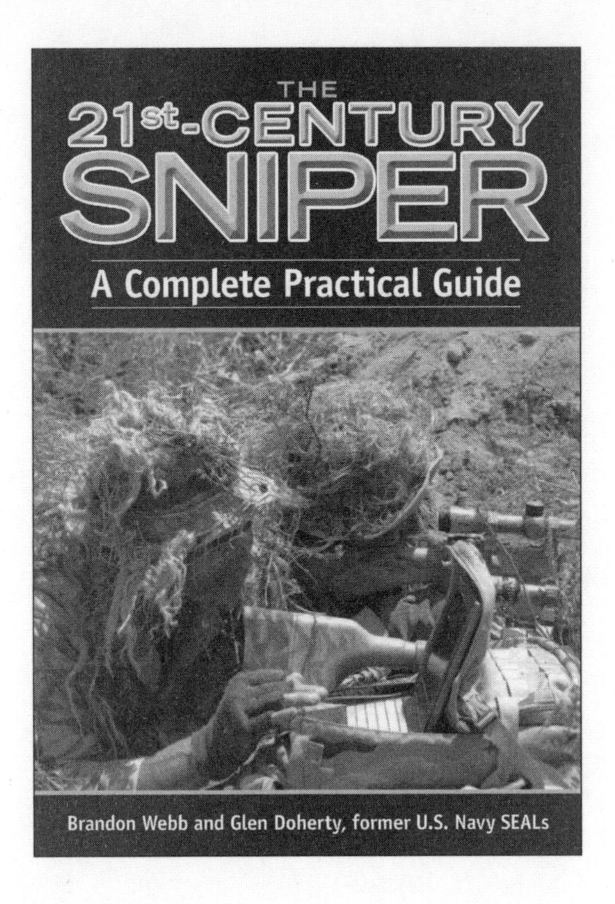

The 21st-Century Sniper

A Complete Practical Guide

Brandon Webb and Glen Doherty, former U.S. Navy SEALS

The twenty-first-century sniper is a mature, intelligent shooter who leverages technology to his deadly advantage. He has spent thousands of hours honing his skills. He is a master of concealment in all environments, from the mountains of Afghanistan to the crowded streets of Iraq. He is trained in science and left alone to create the unique art of the kill. It is his job to simultaneously utilize tools, training, and creativity to deliver devastating psychological impact upon the battlefield. And it is he alone who is left with the intimacy of the kill.

In this complete practical guide for any modern sniper, former Navy SEAL and military sniper Brandon Webb reveals the tips and basic training necessary to become an efficient marksman. Webb is an international authority on sniping, and after serving multiple missions in Iraq and Afghanistan, he ran the Navy SEAL sniper course, which is arguably the best sniper qualification course in the world. Including details on advanced sniper training for maritime, helicopter, and urban sniper operations, *The 21st-century Sniper* touches on the latest research, development, testing, and evaluation of sniper weapons systems and optics. From trajectories and wind speed to camouflage and best vantage points and targets, Webb covers everything an expert sniper needs to know. This book is suitable for gun enthusiasts, outdoorsmen, the beginning sniper, and those with military backgrounds.

$17.95 Paperback • 320 pages